Praxis Biology

0231
0233
0235

Teacher Certification Exam

By: Sharon Wynne, M.S.

XAMonline, INC.
Boston

Copyright © 2013 XAMonline, Inc.
All rights reserved. No part of the material protected by this copyright notice may be reproduced or utilized in any form or by any means, electronic or mechanical, including photocopying, recording or by any information storage and retrievable system, without written permission from the copyright holder.

To obtain permission(s) to use the material from this work for any purpose including workshops or seminars, please submit a written request to:

XAMonline, Inc.
25 First Street, Suite 106
Cambridge, MA 02141
Toll Free: 1-800-301-4647
Email: info@xamonline.com
Web: www.xamonline.com
Fax: 1-617-583-5552

Library of Congress Cataloging-in-Publication Data

Wynne, Sharon A.
    Biology 0231, 0233, 0235: Teacher Certification / Sharon A. Wynne. -4th ed.
    ISBN 978-1-60787-346-4
    1. Biology 0231, 0232, 0235,   2. Study Guides.   3. Praxis
    4. Teachers' Certification & Licensure.   5. Careers

Disclaimer:
The material presented in this publication is the sole work of XAMonline and was created independently from the National Education Association, Educational Testing Service, or any State Department of Education, National Evaluation Systems or other testing affiliates.

Between the time of publication and printing, state specific standards, testing formats, and website information may change. XAMonline developed the sample test questions and they reflect similar content as on real tests; however, they are not former tests. XAMonline assembles content that aligns with state standards, but makes no claims or guarantees regarding test performance. Numerical scores are determined by testing companies such as NES or ETS and then are compared with individual state standards. A passing score varies from state to state.

Printed in the United States of America

Praxis Biology 0231, 0233, 0235
ISBN: 978-1-60787-346-4

TEACHER CERTIFICATION STUDY GUIDE

Table of Contents

DOMAIN 1.0   BASIC PRINCIPLES OF SCIENCE (HISTORY AND NATURAL SCIENCE)

Competency 1.1   Nature of scientific knowledge, inquiry, and historical perspectives

Skill 1.1.1   Scientific methods 1

Skill 1.1.2   Processes involved in scientific inquiry 2

Skill 1.1.3   Process skills 2

Skill 1.1.4   Facts 3

Skill 1.1.5   Concepts 3

Skill 1.1.6   Models 4

Skill 1.1.7   Commonly shared scientific ideals 4

Skill 1.1.8   Philosophy 6

Skill 1.1.9   Contributions made by major historical figures and landmark events in the field of biology 6

Competency 1.2   Mathematics, measurement, and data manipulation

Skill 1.2.1   Measurement and notation systems 8

Skill 1.2.2   Data collection 9

Skill 1.2.3   Data manipulation 9

Skill 1.2.4   Data interpretation 11

Skill 1.2.5   Data presentation (tables, graphs, charts, error analysis) 12

Competency 1.3   Laboratory procedures and safety

Skill 1.3.1   Safe preparation, storage, use, and disposal of laboratory and field materials 13

Skill 1.3.2   Selection and use of appropriate laboratory equipment 14

TEACHER CERTIFICATION STUDY GUIDE

Skill 1.3.3 Safety and emergency procedures of the science classroom and laboratory 16

Skill 1.3.4 Legal responsibilities of the teacher 17

DOMAIN 2.0   MOLECULAR AND CELLULAR BIOLOGY

Competency 2.1 Chemical basis of life

Skill 2.1.1 Basic chemical structures 18

Skill 2.1.2 Atoms, molecules, and chemical bonds 19

Skill 2.1.3 pH and buffers 21

Skill 2.1.4 Biologically important molecules 22

Skill 2.1.5 Thermodynamics and free energy 25

Skill 2.1.6 Cellular bioenergetics 27

Skill 2.1.7 Photosynthesis 28

Skill 2.1.8 Respiration 31

Skill 2.1.9 Enzymes 35

Competency 2.2 Cell Structure and function

Skill 2.2.1 Membranes, organelles, and subcellular components of cells 37

Skill 2.2.2 Cell cycle, cytokinesis, mitosis, and meiosis 48

Competency 2.3 Molecular basis of heredity

Skill 2.3.1 Structure and function of nucleic acids 49

Skill 2.3.2 DNA replication 50

Skill 2.3.3 Protein synthesis 53

Skill 2.3.4 Gene regulation 55

Skill 2.3.5 Mutation and transposable elements 56

Skill 2.3.6 Viruses 57

TEACHER CERTIFICATION STUDY GUIDE

Skill 2.3.7 Molecular basis of genetic diseases (cancer, sickle-cell anemia, hemophilia)..57

Skill 2.3.8 Recombinant DNA and genetic engineering................................59

Skill 2.3.9 Genome mapping of humans and other organisms....................62

DOMAIN 3.0   CLASSICAL GENETICS AND EVOLUTION

Competency 3.1 Classical Genetics

Skill 3.1.1 Mendelian inheritance and probability 63

Skill 3.1.2 Non-Mendelian inheritance...66

Skill 3.1.3 Linkage...67

Skill 3.1.4 Human genetic disorders..67

Skill 3.1.5 Interaction between heredity and the environment......................69

Competency 3.2 Evolution

Skill 3.2.1 Evidence...71

Skill 3.2.2 Mechanisms 74

Skill 3.2.3 Population genetics 78

Skill 3.2.4 Speciation...82

Skill 3.2.5 Phylogeny...84

Skill 3.2.6 Origin of life 85

Skill 3.2.7 Species extinction...86

DOMAIN 4.0   DIVERSITY OF LIFE, PLANTS, AND ANIMALS

Competency 4.1 Diversity of life

Skill 4.1.1 Classification schemes and the five-kingdom system..................88

Skill 4.1.2 Characteristics and representatives of kingdoms 90

BIOLOGY

TEACHER CERTIFICATION STUDY GUIDE

Competency 4.2 Plants

Skill 4.2.1 Evolution (including adaptation to land and major divisions) 96

Skill 4.2.2 Anatomy (including roots, stems, leaves, and reproductive structures) 98

Skill 4.2.3 Physiology (C3 and C4 photosynthesis, hormones, photoperiods, water and nutrient uptake, translocation).................................. 101

Skill 4.2.4 Reproduction (alternation of generations, fertilization and zygote formation, dispersal, germination, growth and differentiation, vegetative propagation) 105

Competency 4.3 Animals

Skill 4.3.1 Evolution (phylogeny and classification, major phyla)...............108

Skill 4.3.2 Life functions and associated structures (digestion, excretion, nervous control, contractile systems and movement, support, integument, immunity, endocrine system)............................110

Skill 4.3.3 Reproduction and development (gametogenesis, fertilization, parthenogenesis, embryogenesis, growth and differentiation, metamorphosis, aging)...127

Skill 4.3.4 Behavior (taxes, instincts, learned behaviors, communication)...135

DOMAIN 5.0 ECOLOGY

Competency 5.1 Populations (intraspecific competition, density factors, population growth, dispersion patterns, life-history patterns, social behavior)...136

Competency 5.2 Communities (niche, interspecific relationships, species diversity, succession)..140

Competency 5.3 Ecosystems (terrestrial, aquatic, biomes, energy flow, biogeochemical cycles, stability and disturbances, human impact, interrelationships among ecosystems)..............................145

DOMAIN 6.0 SCIENCE, TECHNOLOGY, AND SOCIETY

Competency 6.1 Impact of science and technology on the environment and Human affairs 156

TEACHER CERTIFICATION STUDY GUIDE

Competency 6.2 Hazards induced by humans and/or nature 157

Competency 6.3 Issues and applications (production, storage, use, management, and disposal of consumer products and energy, and management of natural resources) 158

Competency 6.4 Social, political, ethical, and economic issues in biology 160

Competency 6.5 Societal issues with health and medical advances 162

Sample Essay 164

Sample Test 172

Answer Key 192

Rigor Table 193

Rationale 194

About XAMonline

XAMonline – A Specialty Teacher Certification Company
Created in 1996, XAMonline was the first company to publish study guides for state-specific teacher certification examinations. Founder Sharon Wynne found it frustrating that materials were not available for teacher certification preparation and decided to create the first single, state-specific guide. XAMonline has grown into a company of over 1800 contributors and writers and offers over 300 titles for the entire PRAXIS series and every state examination. No matter what state you plan on teaching in, XAMonline has a unique teacher certification study guide just for you.

XAMonline – Value and Innovation
We are committed to providing value and innovation. Our print-on-demand technology allows us to be the first in the market to reflect changes in test standards and user feedback as they occur. Our guides are written by experienced teachers who are experts in their fields. And, our content reflects the highest standards of quality. Comprehensive practice tests with varied levels of rigor means that your study experience will closely match the actual in-test experience.

To date, XAMonline has helped nearly 600,000 teachers pass their certification or licensing exams. Our commitment to preparation exceeds simply providing the proper material for study - it extends to helping teachers **gain mastery** of the subject matter, giving them the **tools** to become the most effective classroom leaders possible, and ushering today's students toward a **successful future**.

TEACHER CERTIFICATION STUDY GUIDE

About this Study Guide

Purpose of this Guide
Is there a little voice inside of you saying, "Am I ready?" Our goal is to replace that little voice and remove all doubt with a new voice that says, "I AM READY. **Bring it on!**" by offering the highest quality of teacher certification study guides.

Organization of Content
You will see that while every test may start with overlapping general topics, each are very unique in the skills they wish to test. Only XAMonline presents custom content that analyzes deeper than a title, a subarea, or an objective. Only XAMonline presents content and sample test assessments along with **focus statements**, the deepest-level rationale and interpretation of the skills that are unique to the exam.

Title and field number of test
→Each exam has its own name and number. XAMonline's guides are written to give you the content you need to know for the <u>specific</u> exam you are taking. You can be confident when you buy our guide that it contains the information you need to study for the specific test you are taking.

Subareas
→These are the major content categories found on the exam. XAMonline's guides are written to cover all of the subareas found in the test frameworks developed for the exam.

Objectives
→These are standards that are unique to the exam and represent the main subcategories of the subareas/content categories. XAMonline's guides are written to address every specific objective required to pass the exam.

Focus statements
→These are examples and interpretations of the objectives. You find them in parenthesis directly following the objective. They provide detailed examples of the range, type, and level of content that appear on the test questions. **Only XAMonline's guides drill down to this level.**

How do We Compare with Our Competitors?
XAMonline – drills down to the focus statement level
CliffsNotes and REA – organized at the objective level
Kaplan – provides only links to content
MoMedia – content not specific to the test

Each subarea is divided into manageable sections that cover the specific skill areas. Explanations are easy-to-understand and thorough. You'll find that every test answer contains a rejoinder so if you need a refresher or further review after taking the test, you'll know exactly to which section you must return.

TEACHER CERTIFICATION STUDY GUIDE

How to Use this Book
Our informal polls show that most people begin studying up to 8 weeks prior to the test date, so start early. Then ask yourself some questions: How much do you really know? Are you coming to the test straight from your teacher-education program or are you having to review subjects you haven't considered in 10 years? Either way, take a **diagnostic or assessment test** first. Also, spend time on sample tests so that you become accustomed to the way the actual test will appear.

This guide comes with an online diagnostic test of 30 questions found online at www.XAMonline.com. It is a little boot camp to get you up for the task and reveal things about your compendium of knowledge in general. Although this guide is structured to follow the order of the test, you are not required to study in that order. By finding a time-management and study plan that fits your life you will be more effective. The results of your diagnostic or self-assessment test can be a guide for how to manage your time and point you towards an area that needs more attention.

After taking the diagnostic exam, review the competencies and skills covered in the guide and decide which sections you need more work in. Taking this step will give you a study plan for each chapter.

Week	Activity
8 weeks prior to test	Take a diagnostic test found at www.XAMonline.com
7 weeks prior to test	Review the competencies and skills covered in the guide and decide which sections you need to focus on and which you can skip.
6-3 weeks prior to test	For each of these 4 weeks, choose a content area to study. You don't have to go in the order of the book. It may be that you start with the content that needs the most review. Alternately, you may want to ease yourself into plan by starting with the most familiar material.
2 weeks prior to test	Take the sample test, score it, and create a review plan for the final week before the test.
1 week prior to test	Following your plan (which will likely be aligned with the areas that need the most review) go back and study the sections that align with the questions you may have gotten wrong. Then go back and study the sections related to the questions you answered correctly. If need be, create flashcards and drill yourself on any area that you makes you anxious.

TEACHER CERTIFICATION STUDY GUIDE

About the PRAXIS Exams

What is PRAXIS?
PRAXIS II tests measure the knowledge of specific content areas in K-12 education. The test is a way of insuring that educators are prepared to not only teach in a particular subject area, but also have the necessary teaching skills to be effective. The Educational Testing Service administers the test in most states and has worked with the states to develop the material so that it is appropriate for state standards.

PRAXIS Points
1. The PRAXIS Series comprises more than 140 different tests in over 70 different subject areas.
2. Over 90% of the PRAXIS tests measure subject area knowledge.
3. The purpose of the test is to measure whether the teacher candidate possesses a sufficient level of knowledge and skills to perform job duties effectively and responsibly.
4. Your state sets the acceptable passing score.
5. Any candidate, whether from a traditional teaching-preparation path or an alternative route, can seek to enter the teaching profession by taking a PRAXIS test.
6. PRAXIS tests are updated regularly to ensure current content.

Often **your own state's requirements** determine whether or not you should take any particular test. The most reliable source of information regarding this is either your state's Department of Education or the Educational Testing Service. Either resource should also have a complete list of testing centers and dates. Test dates vary by subject area and not all test dates necessarily include your particular test, so be sure to check carefully.

If you are in a teacher-education program, check with the Education Department or the Certification Officer for specific information for testing and testing timelines. The Certification Office should have most of the information you need.

If you choose an alternative route to certification you can either rely on our website at www.XAMonline.com or on the resources provided by an alternative certification program. Many states now have specific agencies devoted to alternative certification and there are some national organizations as well:
National Center for Education Information
http://www.ncei.com/Alt-Teacher-Cert.htm
National Associate for Alternative Certification
http://www.alt-teachercert.org/index.asp

Interpreting Test Results
Contrary to what you may have heard, the results of a PRAXIS test are not based on time. More accurately, you will be scored on the raw number of points

TEACHER CERTIFICATION STUDY GUIDE

you earn in relation to the raw number of points available. Each question is worth one raw point. It is likely to your benefit to complete as many questions in the time allotted, but it will not necessarily work to your advantage if you hurry through the test.

Follow the guidelines provided by ETS for interpreting your score. The web site offers a sample test score sheet and clearly explains how the scores are scaled and what to expect if you have an essay portion on your test.

Scores are usually available by phone within a month of the test date and scores will be sent to your chosen institution(s) within six weeks. Additionally, ETS now makes online, downloadable reports available for 45 days from the reporting date.

It is **critical** that you be aware of your own state's passing score. Your raw score may qualify you to teach in some states, but not all. ETS administers the test and assigns a score, but the states make their own interpretations and, in some cases, consider combined scores if you are testing in more than one area.

What's on the Test?
PRAXIS tests vary from subject to subject and sometimes even within subject area. For PRAXIS Biology, the 0231 test lasts for 1 hour and consists of approximately 75 multiple-choice questions; the 0235 test lasts for 2 hours and consists of approximately 150 multiple-choice questions; the 0233 test lasts for 1 hour and consists of 3 constructed response essays. The breakdown of the questions is as follows:

Category	Approximate Number of Questions	Approximate Percentage of the test
0231 and 0235 Content Knowledge		
I: Basic Principles of Science	0231: 13 0235: 12	0231: 17% 0235: 8%
II: Molecular and Cellular Biology	0231: 12 0235: 38	0231: 16% 0235: 25%
III: Classical Genetics and Evolution	0231: 11 0235: 23	0231: 15% 0235: 15%
IV: Diversity of Life, Plants, and Animals	0231: 19 0235: 45	0231: 26% 0235: 30%
V: Ecology	0231: 10 0235: 22	0231: 13% 0235: 15%
VI: Science, Technology, and Society	0231: 10 0235: 10	0231: 13% 0235: 7%
0233 Content Essays		
I: Molecular and Cellular Biology	1	33.3%
II: Classical Genetics and Evolution	1	33.3%
III: Organismal Biology and Ecology	1	33.3%

TEACHER CERTIFICATION STUDY GUIDE

Question Types
You're probably thinking, enough already, I want to study! Indulge us a little longer while we explain that there is actually more than one type of multiple-choice question. You can thank us later after you realize how well prepared you are for your exam.

1. **Complete the Statement.** The name says it all. In this question type you'll be asked to choose the correct completion of a given statement. For example: The Dolch Basic Sight Words consist of a relatively short list of words that children should be able to:
   a. Sound out
   b. Know the meaning of
   c. Recognize on sight
   d. Use in a sentence

   The correct answer is A. In order to check your answer, test out the statement by adding the choices to the end of it.

2. **Which of the Following.** One way to test your answer choice for this type of question is to replace the phrase "which of the following" with your selection. Use this example: Which of the following words is one of the twelve most frequently used in children's reading texts:
   a. There
   b. This
   c. The
   d. An

   Don't look! Test your answer. ____ is one of the twelve most frequently used in children's reading texts. Did you guess C? Then you guessed correctly.

3. **Roman Numeral Choices.** This question type is used when there is more than one possible correct answer. For example: Which of the following two arguments accurately supports the use of cooperative learning as an effective method of instruction?
   I. Cooperative learning groups facilitate healthy competition between individuals in the group.
   II. Cooperative learning groups allow academic achievers to carry or cover for academic underachievers.
   III. Cooperative learning groups make each student in the group accountable for the success of the group.
   IV. Cooperative learning groups make it possible for students to reward other group members for achieving.

   A. I and II
   B. II and III
   C. I and III
   D. III and IV

BIOLOGY

Notice that the question states there are **two** possible answers. It's best to read all the possibilities first before looking at the answer choices. In this case, the correct answer is D.

4. **Negative Questions.** This type of question contains words such as "not," "least," and "except." Each correct answer will be the statement that does **not** fit the situation described in the question. Such as: Multicultural education is **not**
    a. An idea or concept
    b. A "tack-on" to the school curriculum
    c. An educational reform movement
    d. A process

    Think to yourself that the statement could be anything but the correct answer. This question form is more open to interpretation than other types, so read carefully and don't forget that you're answering a negative statement.

5. **Questions That Include Graphs, Tables, or Reading Passages.** As ever, read the question carefully. It likely asks for a very specific answer and not broad interpretation of the visual. Here is a simple (though not statistically accurate) example of a graph question: In the following graph in how many years did more men take the NYSTCE exam than women?

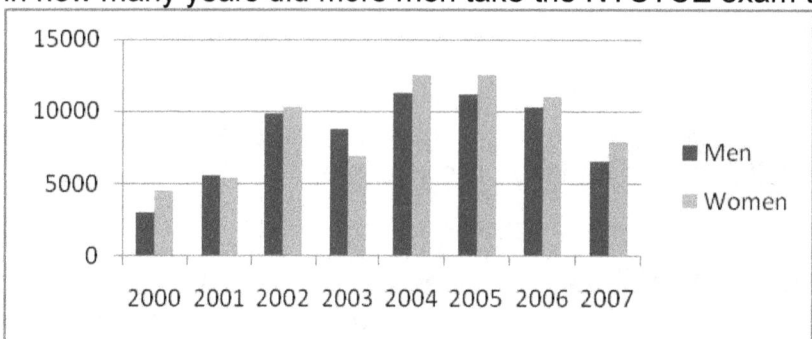

    a. None
    b. One
    c. Two
    d. Three

    It may help you to simply circle the two years that answer the question. Make sure you've read the question thoroughly and once you've made your determination, double check your work. The correct answer is C.

TEACHER CERTIFICATION STUDY GUIDE

Helpful Hints

Study Tips
1. **You are what you eat.** Certain foods aid the learning process by releasing natural memory enhancers called CCKs (cholecystokinin) composed of tryptophan, choline, and phenylalanine. All of these chemicals enhance the neurotransmitters associated with memory and certain foods release memory enhancing chemicals. A light meal or snacks from the following foods fall into this category:
   - Milk
   - Nuts and seeds
   - Rice
   - Oats
   - Eggs
   - Turkey
   - Fish

   The better the connections, the more you comprehend!

2. **See the forest for the trees.** In other words, get the concept before you look at the details. One way to do this is to take notes as you read, paraphrasing or summarizing in your own words. Putting the concept in terms that are comfortable and familiar may increase retention.

3. **Question authority.** Ask why, why, why. Pull apart written material paragraph by paragraph and don't forget the captions under the illustrations. For example, if a heading reads *Stream Erosion* put it in the form of a question (why do streams erode? Or what is stream erosion?) then find the answer within the material. If you train your mind to think in this manner you will learn more and prepare yourself for answering test questions.

4. **Play mind games.** Using your brain for reading or puzzles keeps it flexible. Even with a limited amount of time your brain can take in data (much like a computer) and store it for later use. In ten minutes you can: read two paragraphs (at least), quiz yourself with flash cards, or review notes. Even if you don't fully understand something on the first pass, your mind stores it for recall, which is why frequent reading or review increases chances of retention and comprehension.

5. **The pen is mightier than the sword.** Learn to take great notes. A by-product of our modern culture is that we have grown accustomed to getting our information in short doses. We've subconsciously trained ourselves to assimilate information into neat little packages. Messy notes fragment the flow of information. Your notes can be much clearer with proper formatting. **The Cornell Method** is one such format. This method was popularized in *How to Study in College,* Ninth Edition, by Walter

BIOLOGY

Pauk. You can benefit from the method without purchasing an additional book by simply looking the method up online. Below is a sample of how *The Cornell Method* can be adapted for use with this guide.

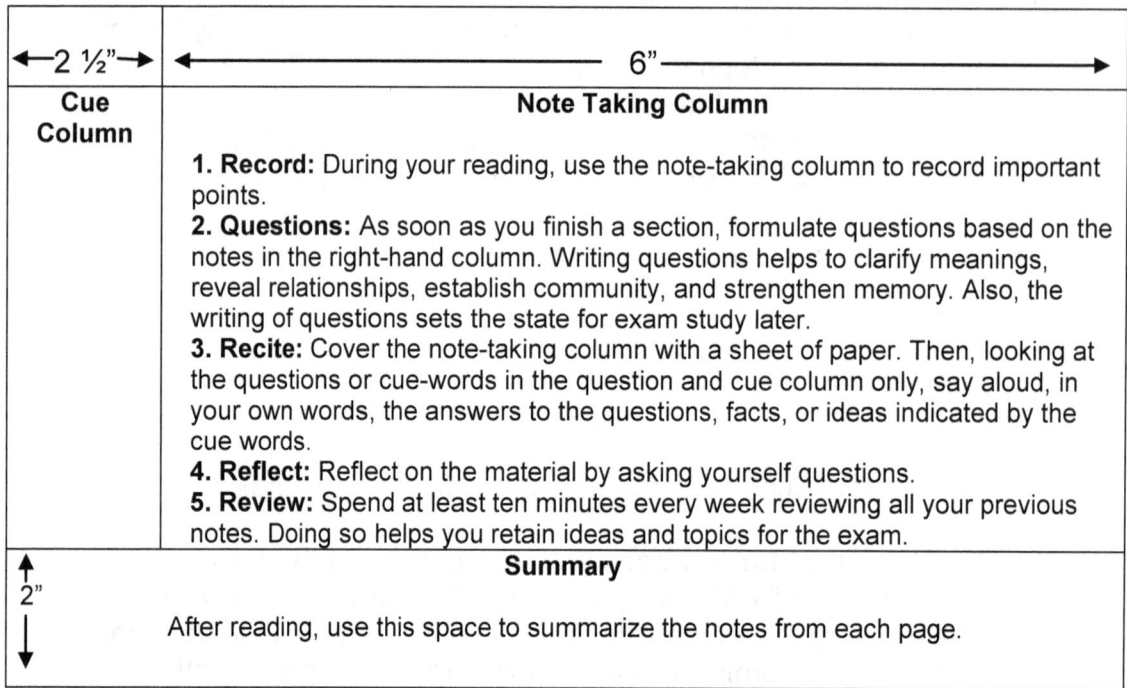

*Adapted from *How to Study in College,* Ninth Edition, by Walter Pauk, ©2008 Wadsworth

6. **Place yourself in exile and set the mood.** Set aside a particular place and time to study that best suits your personal needs and biorhythms. If you're a night person, burn the midnight oil. If you're a morning person set yourself up with some coffee and get to it. Make your study time and place as free from distraction as possible and surround yourself with what you need, be it silence or music. Studies have shown that music can aid in concentration, absorption, and retrieval of information. Not all music, though. Classical music is said to work best.

7. **Get pointed in the right direction.** Use arrows to point to important passages or pieces of information. It's easier to read than a page full of yellow highlights. Highlighting can be used sparingly, but add an arrow to the margin to call attention to it.

8. **Check your budget.** You should at least review all the content material before your test, but allocate the most amount of time to the areas that need the most refreshing. It sounds obvious, but it's easy to forget. You can use the study rubric above to balance your study budget.

> The proctor will write the start time where it can be seen and then, later, provide the time remaining, typically 15 minutes before the end of the test.

TEACHER CERTIFICATION STUDY GUIDE

Testing Tips

1. **Get smart, play dumb.** Sometimes a question is just a question. No one is out to trick you, so don't assume that the test writer is looking for something other than what was asked. Stick to the question as written and don't overanalyze.

2. **Do a double take.** Read test questions and answer choices at least twice because it's easy to miss something, to transpose a word or some letters. If you have no idea what the correct answer is, skip it and come back later if there's time. If you're still clueless, it's okay to guess. Remember, you're scored on the number of questions you answer correctly and you're not penalized for wrong answers. The worst case scenario is that you miss a point from a good guess.

3. **Turn it on its ear.** The syntax of a question can often provide a clue, so make things interesting and turn the question into a statement to see if it changes the meaning or relates better (or worse) to the answer choices.

4. **Get out your magnifying glass.** Look for hidden clues in the questions because it's difficult to write a multiple-choice question without giving away part of the answer in the options presented. In most questions you can readily eliminate one or two potential answers, increasing your chances of answering correctly to 50/50, which will help out if you've skipped a question and gone back to it (see tip #2).

5. **Call it intuition.** Often your first instinct is correct. If you've been studying the content you've likely absorbed something and have subconsciously retained the knowledge. On questions you're not sure about trust your instincts because a first impression is usually correct.

6. **Graffiti.** Sometimes it's a good idea to mark your answers directly on the test booklet and go back to fill in the optical scan sheet later. You don't get extra points for perfectly blackened ovals. If you choose to manage your test this way, be sure not to mismark your answers when you transcribe to the scan sheet.

7. **Become a clock-watcher.** You have a set amount of time to answer the questions. Don't get bogged down laboring over a question you're not sure about when there are ten others you could answer more readily. If you choose to follow the advice of tip #6, be sure you leave time near the end to go back and fill in the scan sheet.

BIOLOGY

TEACHER CERTIFICATION STUDY GUIDE

Do the Drill

No matter how prepared you feel it's sometimes a good idea to apply Murphy's Law. So the following tips might seem silly, mundane, or obvious, but we're including them anyway.

1. **Remember, you are what you eat, so bring a snack.** Choose from the list of energizing foods that appear earlier in the introduction.
2. **You're not too sexy for your test.** Wear comfortable clothes. You'll be distracted if your belt is too tight, or if you're too cold or too hot.
3. **Lie to yourself.** Even if you think you're a prompt person, pretend you're not and leave plenty of time to get to the testing center. Map it out ahead of time and do a dry run if you have to. There's no need to add road rage to your list of anxieties.
4. **Bring sharp, number 2 pencils.** It may seem impossible to forget this need from your school days, but you might. And make sure the erasers are intact, too.
5. **No ticket, no test.** Bring your admission ticket as well as **two** forms of identification, including one with a picture and signature. You will not be admitted to the test without these things.
6. **You can't take it with you**. Leave any study aids, dictionaries, notebooks, computers and the like at home. Certain tests **do** allow a scientific or four-function calculator, so check ahead of time if your test does.
7. **Prepare for the desert**. Any time spent on a bathroom break **cannot** be made up later, so use your judgment on the amount you eat or drink.
8. **Quiet, Please!** Keeping your own time is a good idea, but not with a timepiece that has a loud ticker. If you use a watch, take it off and place it nearby but not so that it distracts you. And **silence your cell phone.**

To the best of our ability, we have compiled the content you need to know in this book and in the accompanying online resources. The rest is up to you. You can use the study and testing tips or you can follow your own methods. Either way, you can be confident that there aren't any missing pieces of information and there shouldn't be any surprises in the content on the test.

If you have questions about test fees, registration, electronic testing, or other content verification issues please visit www.ets.org.

Good luck!
Sharon Wynne
Founder, XAMonline

DOMAIN 1.0 BASIC PRINCIPLES OF SCIENCE (HISTORY AND NATURAL SCIENCE)

Competency 1.1 Nature of scientific knowledge, inquiry, and historical perspectives:

Skill 1.1.1 Scientific methods

Scientific research serves two purposes –
1. To investigate and acquire knowledge that is theoretical *and*
2. To perform research which is of practical value

Science has the unique ability to serve humanity. Scientific research results from inquiry. An inquiring mind is thirsty, trying to find answers. An inquisitive person asks questions and wants to find answers. The two most important questions – why and how – are the starting points of all scientific inquiry.

Scientific research uses the scientific method to answer questions. Researchers follow the scientific method, which consists of a series of steps designed to solve a problem.

The aim of the scientific method is to eliminate any bias or prejudice that the scientist or researcher may bring to the inquiry. If we follow all the steps of the scientific method as outlined, we achieve maximum elimination of bias.

The scientific method consists of the following steps:

1. Stating the problem clearly and precisely
2. Gathering information/research
3. Formulating a hypothesis (an educated guess)
4. Designing an experiment
5. Analyzing the results
6. Drawing a conclusion

TEACHER CERTIFICATION STUDY GUIDE

Sample Test Question and Rationale

1. Identify the control in the following experiment. A student grew four plants under the following conditions and measured photosynthetic rate by measuring mass. He grew two plants in 50% light and two plants in 100% light.
(Average Rigor)

   A. plants grown with no added nutrients.
   B. plants grown in the dark.
   C. plants grown in 100% light.
   D. plants grown in 50% light.

Answer: C. plants grown in 100% light.

The plants grown in 100% light are the control that the student will compare the plants grown in 50% light.

Skill 1.1.2   Processes involved in scientific inquiry

Science is a body of knowledge systematically derived from study, observations, and experimentation. The objective of science is to identify and establish principles and theories that are utilized to solve problems. Pseudoscience, on the other hand, involves beliefs that are not supported by hard evidence. In other words, there is no scientific methodology or application. Some classic examples of pseudoscience include witchcraft, alien encounters, or any topic explained by hearsay.

Scientific experimentation must be repeatable. Experimentation results in theories that can be disproved and changed. Science depends on communication, agreement, and disagreement among scientists. It is composed of theories, laws, and hypotheses.

Skill 1.1.3   Process skills

Theory - A statement of principles or relationships relating to a natural event or phenomenon, which has been verified and accepted.

Law - An explanation of events that occur with uniformity under the same conditions (e.g., laws of nature, law of gravitation).

Hypothesis - An unproved theory or educated guess followed by research to best explain a phenomenon. A theory is a proven hypothesis.

TEACHER CERTIFICATION STUDY GUIDE

Science is limited by the available technology. An example of this would be the relationship between the discovery of the cell and the invention of the microscope. As our technology improves, more hypotheses will become theories and possibly laws. Data collection methods also limit scientific inquiry. Data may be interpreted differently on different occasions. The inherent limitations of scientific methodology produce results or explanations that are subject to change as new technologies emerge.

Skill 1.1.4 Facts

Facts are not always as finite as they appear. More commonly in science, information is a hypothesis or, once tested and confirmed, a theory. Theories exist for long periods and repeatedly receive challenges. Only when a theory has withstood every challenge and been proven to provide reproducible results does it become recognized as a law. It is the universal recognition that defines a theory as a scientific law.

Skill 1.1.5 Concepts

A concept is a general understanding or belief. Scientists challenge concepts. The purpose of the scientific method is to derive clear, unbiased data. Concepts, on the other hand, may be fraught with personal biases and gray areas, overly simplistic, or too encompassing. A scientist might examine a concept, and then try to confirm it by making and testing a hypothesis. In this way, scientific inquiry is more specific and concepts are more generalized.

Sample Test Question and Rationale

2. **The concept that the rate of a given process is controlled by the most scarce factor in the process is known as?**
   (Average Rigor)

   A. The Rate of Origination.
   B. The Law of the Minimum.
   C. The Law of Limitation.
   D. The Law of Conservation.

Answer: B. The Law of the Minimum

A limiting factor is the component of a biological process that determines how quickly or slowly the process proceeds. Photosynthesis is the main biological process determining the rate of ecosystem productivity or the rate at which an ecosystem creates biomass. Thus, in evaluating the productivity of an ecosystem, potential limiting factors are light intensity, gas concentrations, and mineral availability. The Law of the Minimum states that the required factor which is most scarce in a given process controls the rate of the process.

Skill 1.1.6 Models

Models are the basis for greater understanding. Models are usually small-scale representations that help us understand a larger system. Models aid us by making unusually large or small items more concrete. Common models include the solar system and the DNA helix. It is important to note that models are created with information. How current and accurate the information is at the time of creation may make the model more or less useful later. For example, although Pluto has been considered a planet for many years, it is now considered a dwarf planet. This is due to the progressive nature of science; the more we learn, the more we are forced to reevaluate.

Skill 1.1.7 Commonly shared scientific ideals

Biological science is closely connected to other scientific disciplines and technology resulting in a tremendous impact on society and everyday life. Scientific discoveries often lead to technological advances. Conversely, technology is often necessary for scientific investigation and advances in technology often expand the reach of scientific discoveries. In addition, biology and the other scientific disciplines share severa concepts and processes that help unify the study of science. Finally, because biology is the science of living systems, biology directly affects society and everyday life.

Science and technology, while distinct concepts, are closely related. Science attempts to investigate and explain the natural world, while technology attempts to solve human adaptation problems. Technology often results from the application of scientific discoveries, and advances in technology can increase the impact of scientific discoveries. For example, Watson and Crick used science to discover the structure of DNA and their discovery led to many biotechnological advances in the field of genomics. These technological advances greatly influenced the medical and pharmaceutical fields. The success of Watson and Crick's experiments, however, was dependent on the technology available. Without the necessary technology, the experiments would have failed.

The combination of biology and technology has improved the human standard of living in many ways. However, the negative impact of increasing human life expectancy and population on the environment is problematic. In addition, advances in biotechnology (e.g. genetic engineering, cloning) produce ethical dilemmas that society must consider.

TEACHER CERTIFICATION STUDY GUIDE

The following are the concepts and processes generally recognized as common to all scientific disciplines:

- Systems, order, and organization
- Evidence, models, and explanation
- Constancy, change, and measurement
- Evolution and equilibrium
- Form and function

Because the natural world is so complex, the study of science involves the **organization** of items into smaller groups based on interaction or interdependence. These groups are called **systems**. Examples of organization are the periodic table of elements and the five-kingdom classification scheme for living organisms.

Examples of systems are the solar system, cardiovascular system, Newton's laws of force and motion, and the laws of conservation.

Order refers to the behavior and measurability of organisms and events in nature. The arrangement of planets in the solar system and the life cycle of bacterial cells are examples of order.

Scientists use **evidence** and **models** to form **explanations** of natural events. Models are miniaturized representations of a larger event or system. Evidence is anything that furnishes proof.

Constancy and **change** describe the observable properties of natural organisms and events. Scientists use different systems of **measurement** to observe change and constancy. For example, the freezing and melting point of a given substance and the speed of sound are constant under constant conditions. Growth, decay, and erosion are all examples of natural change.

Evolution is the process of change over a long period of time. While biological evolution is the most common example, one can also classify technological advancement, changes in the universe, and changes in the environment as evolution.

Equilibrium is the state of balance between opposing forces of change. Homeostasis and ecological balance are examples of equilibrium.

Form and **function** are properties of organisms and systems that are closely related. The function of an object usually dictates its form and the form of an object usually facilitates its function. For example, the form of the heart (e.g. muscle, valves) allows it to perform its function of circulating blood through the body.

Skill 1.1.8 Philosophy

To understand scientific ethics, we need to have a clear understanding of general ethics. Ethics is a system of public, general rules that guide human conduct. The rules are general because they apply to all people at all times and they are public because they are not secret codes or practices.

Philosophers have provided a number of moral theories to justify moral rules ranging from utilitarianism to social contract theory. Utilitarianism, proposed by Mozi, a Chinese philosopher who lived from 471-381 BC, is a theory of ethics based on the determination of what is best for the greatest number of people. Kantianism, a theory proposed by Immanuel Kant, a German philosopher who lived from 1724-1804, ascribes intrinsic value to rational beings. Kantianism is the philosophical foundation of contemporary human rights. Social contract theory, on the other hand, is a view of the ancient Greeks, which states that a person's moral and/or political obligations are dependent upon a contract or societal agreement.
The following are some of the guiding principles of scientific ethics:

1. Scientific Honesty: refrain from fabricating or misinterpreting data for personal gain
2. Caution: avoid errors and sloppiness in all scientific experimentation
3. Credit: give credit where credit is due and do not copy
4. Responsibility: report reliable information to the public and do not mislead in the name of science
5. Freedom: freedom to criticize old ideas, question new research, and conduct independent research.

Scientists should show good conduct in their scientific pursuits. Conduct here refers to all aspects of scientific activity including experimentation, testing, education, data evaluation, data analysis, data storing, and peer review.

Skill 1.1.9 Contributions made by major historical figures and landmark events in the field of biology

Anton van Leeuwenhoek is known as the father of microscopy. In the 1650s, Leeuwenhoek began making tiny lenses that produced magnifications up to 300 times. He was the first to see and describe bacteria, yeast, plants, and the microscopic life found in water. Over the years, light microscopes have advanced to produce greater clarity and magnification. The scanning electron microscope (SEM) was developed in the 1950s. Instead of light, a beam of electrons pass through the specimen. Scanning electron microscopes have a resolution about one thousand times greater than light microscopes. The disadvantage of the SEM is that the chemical and physical methods used to prepare the sample result in the death of the specimen.

TEACHER CERTIFICATION STUDY GUIDE

In the late 1800s, Pasteur invented pasteurization, the first rabies vaccine, and performed experiments that supported the germ theory of disease. Koch took this observation one-step further by formulating a theory that specific pathogens cause specific diseases. Scientists still use **Koch's postulates** as guidelines in the field of microbiology. The guidelines state that the same pathogen must be found in every diseased person, the pathogen must be isolated and grown in culture, the pathogen must induce disease in experimental animals, and the same pathogen must be isolated from the experimental animal.

The discovery of the structure of DNA was another key event in biological history. In the 1950s, James Watson and Francis Crick identified the structure of a DNA molecule as that of a double helix. This structure made it possible to explain DNA's ability to replicate and to control the synthesis of proteins.

The use of animals in biological research has expedited many scientific discoveries. Animal research has allowed scientists to learn more about animal biological systems, including the circulatory and reproductive systems. One significant use of animals is for the testing of drugs, vaccines, and other products (such as perfumes and shampoos) before use or consumption by humans.

Sample Test Question and Rationale

3. **Which of the following discovered penicillin?**
   (Rigorous)

   A. Pierre Curie
   B. Becquerel
   C. Louis Pasteur
   D. Alexander Fleming

Answer: D. Alexander Fleming

Sir Alexander Fleming was a pharmacologist from Scotland. He isolated the antibiotic penicillin from a fungus in 1928.

Competency 1.2   Mathematics, measurement, and data manipulation

Math, science, and technology share many common themes. All three use models, diagrams, and graphs to simplify a concept for analysis and interpretation. Patterns observed in these systems lead to predictions based on these observations. Another common theme among these three systems is equilibrium. **Equilibrium** is a state in which forces are balanced, resulting in stability. Static equilibrium is stability due to a lack of changes, and dynamic equilibrium is stability due to a balance between opposing forces.

Skill 1.2.1   Measurement and notation systems

Science uses the **metric system**, as it is accepted worldwide and allows the results of experiments, performed by different scientists around the world, to be compared to one another. The meter is the basic metric unit of length. One meter is 1.1 yards. The liter is the basic metric unit of volume. 3.846 liters is 1 gallon. The gram is the basic metric unit of mass. One thousand grams is 2.2 pounds. The following prefixes define multiples of the basic metric units.

Prefix	Multiplying factor	Prefix	Multiplying factor
deca-	10X the base unit	deci-	1/10 the base unit
hecto-	100X	centi-	1/100
kilo-	1,000X	milli-	1/1,000
mega-	1,000,000X	micro-	1/1,000,000
giga-	1,000,000,000X	nano-	1/1,000,000,000
tera-	1,000,000,000,000X	pico-	1/1,000,000,000,000

The common instrument used for measuring volume is the graduated cylinder. The standard unit of measurement is milliliters (mL). To ensure accurate measurement, it is important to read the liquid in the cylinder at the bottom of the meniscus, the curved surface of the liquid.

The common instrument used in measuring mass is the triple beam balance. The triple beam balance can accurately measure tenths of a gram and can estimate hundredths of a gram.

The ruler and meter stick are the most commonly used instruments for measuring length. As with all scientific measurements, standard units of length are metric.

Sample Test Question and Rationale

4. The International System of Units (SI) measurement for temperature is on the _____ scale.
   (Rigorous)

   A. Celsius
   B. Farenheit
   C. Kelvin
   D. Rankine

Answer: C. Kelvin

Science uses the SI system because of its worldwide acceptance and ease of comparison. The SI scale for measuring temperature is the Kelvin Scale. Science, however, uses the Celsius scale for its ease of use. The answer is (C).

Skill 1.2.2 Data collection

The procedure used to obtain data is important to the outcome. Experiments consist of **controls** and **variables**. A control is the experiment run under normal, non-manipulated conditions. A variable is a factor or condition the scientist manipulates. In biology, the variable may be light, temperature, pH, time, etc. Scientists can use the differences in tested variables to make predictions or form hypotheses. Only one variable should be tested at a time. In other words, one would not alter both the temperature and pH of the experimental subject.

An **independent variable** is one the researcher directly changes or manipulates. This could be the amount of light given to a plant or the temperature at which bacteria is grown. The **dependent variable** is the factor that changes due to the influence of the independent variable.

Skill 1.2.3 Data manipulation

Data manipulation is important to experimental study. Data manipulation begins by altering one variable at a time, and then assessing the results. Are the results similar to the last time? What has changed? Has it improved or worsened? This process is part of the scientific method, where scientists make predictions and then experiment to test validity. Quite often, this process takes many alterations, and manipulating the data and experimental parameters is useful. We are fortunate to have technological advances to aid us in this area. Biologists use a variety of tools and technologies to perform tests, collect and display data, and analyze relationships. Examples of commonly used tools include computer-linked probes, spreadsheets, and graphing calculators.

Biologists use computer-linked probes to measure various environmental factors including temperature, dissolved oxygen, pH, ionic concentration, and pressure. The advantage of computer-linked probes, as compared to more traditional observational tools, is that the probes automatically gather data and present it in an accessible format. This property of computer-linked probes eliminates the need for constant human observation and manipulation.

Biologists use spreadsheets to organize, analyze, and display data. For example, conservation ecologists use spreadsheets to model population growth and development, apply sampling techniques, and create statistical distributions to analyze relationships. Spreadsheet use simplifies data collection and manipulation and allows the presentation of data in a logical and understandable format.

Graphing programs are another technology with many applications to biology. For example, biologists use algebraic functions to analyze growth, development and other natural processes. Graphing programs can manipulate algebraic data and create graphs for analysis and observation. In addition, biologists use the matrix function of graphing programs to model problems in genetics. The use of graphing programs simplifies the creation of graphical displays including histograms, scatter plots, and line graphs. Finally, biologists connect computer-linked probes, used to collect data, to graphing programs to ease the collection, transmission, and analysis of data.

While it is useful to manipulate data in discovery efforts, it is never acceptable to fabricate or falsely advertise your data.

Sample Test Questions and Rationale

5. **In a data set, the value that occurs with the greatest frequency is referred to as the...**
   (Average Rigor)

   A. Mean
   B. Median
   C. Mode
   D. Range

Answer: C. Mode

Mean is the mathematical average of all the items. The median depends on whether the number of items is odd or even. If the number is odd, then the median is the value of the item in the middle. Mode is the value of the item that occurs the most often, if there are not many items. Bimodal is a situation where there are two items with equal frequency. Range is the difference between the maximum and minimum values.

6. Three plants were grown and the following data recorded. Determine the mean growth.
   (*Easy*)

   Plant 1: 10 cm
   Plant 2: 20 cm
   Plant 3: 15 cm

   A. 5 cm
   B. 45 cm
   C. 12 cm
   D. 15 cm

Answer: D. 15 cm

The mean growth is the average of the three growth heights.

$$\frac{10 + 20 + 15}{3} = 15 \text{ cm average height}$$

Skill 1.2.4 Data interpretation

When interpreting data, one must carefully examine all parameters. You may be attempting to interpret your own data, or to understand data you found in a published format. Either way, it is important to think about what you see. In the scientific realm, numbers are stronger than words, so be sure to provide accurate data and examples to support your comments. By using the scientific method, you will be more likely to catch mistakes, correct biases, and obtain accurate results. When assessing experimental data, utilize proper tools and mathematical concepts. Because people often attempt to use scientific evidence in support of political or personal agendas, the ability to evaluate the credibility of scientific claims is a necessary skill in today's society.

In evaluating scientific claims made in the media, public debates, and advertising, one should follow several guidelines. First, scientific, peer-reviewed journals are the most accepted source for information on scientific experiments and studies. One should carefully scrutinize any claim that does not reference peer-reviewed literature. Second, the media and those with an agenda to advance (e.g., advertisers and debaters) often overemphasize the certainty and importance of experimental results. One should question any scientific claim that sounds either too good to be true or overly certain. Finally, knowledge of experimental design and the scientific method is important in evaluating the credibility of studies. For example, one should look for the inclusion of control groups and the presence of data to support the given conclusions.

TEACHER CERTIFICATION STUDY GUIDE

Skill 1.2.5 Data presentation (tables, graphs, charts, error analysis)

The type of graphic representation used to display observations depends on the type of data collected. **Line graphs** compare different sets of related data and help predict data. For example, a line graph could compare the rate of activity of different enzymes at varying temperatures. A **bar graph** or **histogram** compares different items and helps make comparisons based on the data. For example, a bar graph could compare the ages of children in a classroom. A **pie chart** is useful when organizing data as part of a whole. For example, a pie chart could display the percent of time students spend on various after school activities.

As previously noted, the researcher controls the independent variable. The independent variable is placed on the x-axis (horizontal axis). The dependent variable is influenced by the independent variable and is placed on the y-axis (vertical axis). It is important to choose the appropriate units for labeling the axes. It is best to divide the largest value to be plotted by the number of blocks on the graph, and round to the nearest whole number.

Sample Test Question and Rationale

7. In which of the following situations would a linear extrapolation of data be appropriate?
   (Rigorous)

   A. Computing the death rate of an emerging disease
   B. Computing the number of plant species in a forest over time
   C. Computing the rate of diffusion with a constant gradient
   D. Computing the population at equilibrium

Answer: C: Computing the rate of diffusion with a constant gradient.

The individual data points on a linear graph cluster around a line of best fit. In other words, a relationship is linear if we can sketch a straight line that roughly fits the data points. Extrapolation is the process of estimating data points outside a known set of data points. When extrapolating data of a linear relationship, we extend the line of best fit beyond the known values. The extension of the line represents the estimated data points. Extrapolating data is only appropriate if we are relatively certain that the relationship is indeed linear. The answer is (C).

Competency 1.3 Laboratory procedures and safety

Skill 1.3.1 Safe preparation, storage, use, and disposal of laboratory and field materials

All laboratory solutions should be prepared as directed in the lab manual. Care should be taken to avoid contamination. All glassware should be rinsed thoroughly with distilled water before using and cleaned well after use. All solutions should be made with distilled water as tap water contains dissolved particles that may affect the results of an experiment. Unused solutions should be disposed of according to local disposal procedures.

The "Right to Know Law" covers science teachers who work with potentially hazardous chemicals. Briefly, the law states that employees must be informed of potentially toxic chemicals. An inventory must be made available if requested. The inventory must contain information about the hazards and properties of the chemicals. This inventory is to be checked against the "Substance List". Training must be provided on safe handling and interpretation of the Material Safety Data Sheet.

The following chemicals are potential carcinogens and not allowed in school facilities: Acrylonitrile, Arsenic compounds, Asbestos, Benzidine, Benzene, Cadmium compounds, Chloroform, Chromium compounds, Ethylene oxide, Ortho-toluidine, Nickel powder, and Mercury.

Chemicals should not be stored on bench tops or heat sources. They should be stored in groups based on their reactivity with one another and in protective storage cabinets. All containers within the lab must be labeled. Suspected and known carcinogens must be labeled as such and stored in trays to contain leaks and spills.

Chemical waste should be disposed of in properly labeled containers. Waste should be separated based on its reactivity with other chemicals.

Biological material should never be stored near food or water used for human consumption. All biological material should be appropriately labeled. All blood and body fluids should be put in a well-contained container with a secure lid to prevent leaking. All biological waste should be disposed of in biological hazardous waste bags.

Material safety data sheets are available for every chemical and biological substance. These are available directly from the distribution company and the internet. Before using lab equipment, all lab workers should read and understand the equipment manuals.

TEACHER CERTIFICATION STUDY GUIDE

Sample Test Question and Rationale

8. **Which of the following is not usually found on the MSDS for a laboratory chemical?**
   (Rigorous)

   A. Melting Point
   B. Toxicity
   C. Storage Instructions
   D. Cost

Answer: D. Cost

MSDS, or Material Safety Data Sheets, are used to make sure that anyone can easily obtain information about a chemical especially in the event of a spill or accident. This information typically includes physical data, toxicity, health effects, first aid, reactivity, storage, disposal, protective measures, and spill/leak procedures. Cost is not generally included on MSDS's. Costs are generated by the distributor, and seperate suppliers may have different costs. The answer, therefore, is (D).

Skill 1.3.2 Selection and use of appropriate laboratory equipment

Light microscopes are commonly used in high school laboratory experiments. Total magnification is determined by multiplying the magnification of the ocular and objective lenses. Oculars usually magnify 10X and objective lenses usually magnify 10X on low and 40X on high.

Procedures for the care and use of microscopes include:

-cleaning all lenses with lens paper only,
-carrying microscopes with two hands (one on the arm and one on the base),
-always beginning on low power when focusing before switching to high power,
-storing microscopes with the low power objective down,
-always using a coverslip when viewing wet mount slides, *and*
-bringing the objective down to its lowest position then focusing, moving up to avoid breaking the slide or scratching the lens.

Wet mount slides should be made by placing a drop of water on the specimen and then putting a glass coverslip on top of the drop of water. Dropping the coverslip at a forty-five degree angle will help avoid air bubbles.

Chromatography refers to a set of techniques that are used to separate substances based on their different properties such as size or charge. Paper chromatography uses the principles of capillarity to separate substances such as plant pigments. Molecules of a larger size will move more slowly up the paper, whereas smaller molecules will move more quickly producing lines of pigment.

An **indicator** is any substance used to assist in the classification of another substance. An example of an indicator is litmus paper. Litmus paper is a way to measure whether a substance is acidic or basic. Blue litmus turns pink when acid is placed on it and pink litmus turns blue when a base is placed on it. pH paper is a more accurate measure of pH, with the paper turning different colors depending on the pH value.

Spectrophotometers measure the percent of light at different wavelengths absorbed and transmitted by a pigment solution.

Centrifugation involves spinning substances at a high speed. The more dense part of a solution will settle to the bottom of the test tube, where the lighter material will stay on top. Centrifugation is used to separate blood into blood cells and plasma, with the heavier blood cells settling to the bottom.

Electrophoresis uses electrical charges of molecules to separate them according to their size. The molecules, such as DNA or proteins are pulled through a gel towards either the positive end of the gel box (if the material has a negative charge) or the negative end of the gel box (if the material has a positive charge). DNA is negatively charged and moves towards the positive charge. Smaller segments of DNA will move more quickly toward the positive charge than larger segments.

Sample Test Question and Rationale

9. Paper chromatography is most often associated with the separation of...
   (Average Rigor)

   A. nutritional elements.
   B. DNA.
   C. proteins.
   D. plant pigments.

Answer: D. plant pigments.

Paper chromatography uses the principles of capillarity to separate substances such as plant pigments. Molecules of a larger size will move more slowly up the paper, whereas smaller molecules will move more quickly producing lines of pigment.

Skill 1.3.3 Safety and emergency procedures of the science classroom and laboratories

All science labs should contain the following **safety equipment**.

-Fire blanket that is visible and accessible.
-Ground Fault Circuit Interrupters (GFCI) within two feet of water supplies
-Signs designating room exits.
-Emergency shower providing a continuous flow of water.
-Emergency eye wash station that can be activated by the foot or forearm.
-Eye protection for every student.
-A means of sanitizing equipment.
-Emergency exhaust fans providing ventilation to the outside of the building.
-Master cut-off switches for gas, electric, and compressed air. Switches must have permanently attached handles. Cut-off switches must be clearly labeled.
-An ABC fire extinguisher.
-Storage cabinets for flammable materials.
-Chemical spill control kit.
-Fume hood with a motor that is spark proof.
-Protective laboratory aprons made of flame retardant material.
-Signs that will alert of potential hazardous conditions.
-Labeled containers for broken glassware, flammables, corrosives, and waste.

Students should wear safety goggles when performing dissections, heating, or while using acids and bases. Hair should always be tied back and objects should never be placed in the mouth. Food should not be consumed while in the laboratory. Hands should always be washed before and after laboratory experiments. In case of an accident, eye washes and showers should be used for eye contamination or a chemical spill that covers the student's body. Small chemical spills should only be contained and cleaned by the teacher. Kitty litter or a chemical spill kit should be used to clean a spill. For large spills, the school administration and the local fire department should be notified. Biological spills should only be handled by the teacher. Contamination with biological waste can be cleaned by using bleach when appropriate. Accidents and injuries should always be reported to the school administration and local health facilities. The severity of the accident or injury will determine the course of action.

Sample Test Question and Rationale

10. Which item should always be used when using chemicals with noxious vapors?
    (Easy)

    A. eye protection
    B. face shield
    C. fume hood
    D. lab apron

Answer: C. fume hood

Fume hoods are designed to protect the experimenter from chemical fumes. The three other choices do not prevent chemical fumes from entering the respiratory system.

Skill 1.3.4   Legal responsibilities of the teacher

It is the responsibility of the teacher to provide a safe environment for his or her students. Proper supervision greatly reduces the risk of injury and a teacher should never leave a class for any reason without providing alternate supervision. After an accident, two factors are considered, **foreseeability** and **negligence**. Foreseeability is the anticipation that an event may occur under certain circumstances. Negligence is the failure to exercise ordinary or reasonable care. Safety procedures should be a part of the science curriculum and a well-managed classroom is crucial to avoid potential lawsuits.

DOMAIN 2.0 MOLECULAR AND CELLULAR BIOLOGY

Competency 2.1 Chemical basis of life

Skill 2.1.1 Basic chemical structures

An **element** is a substance that cannot be broken down into other substances. Today, scientists have identified 109 elements: 89 are found in nature and 20 are synthetic.

An **atom** is the smallest particle of an element that exhibits the properties of the element. All of the atoms of a particular element are the same. The atoms of each element are different from the atoms of the other elements.

Elements are assigned an identifying symbol of one or two letters. The symbol for oxygen is O and stands for one atom of oxygen. However, because oxygen atoms in nature are joined together in pairs, the symbol O_2 represents oxygen. This pair of oxygen atoms is a molecule. A **molecule** is the smallest particle of a substance that can exist independently and has all of the properties of the substance. A molecule of most elements is made up of one atom. However, oxygen, hydrogen, nitrogen, and chlorine molecules are made of two atoms each.

A **compound** is made of two or more chemically-combined elements. Atoms join when elements are chemically combined. The result is that the elements lose their individual identities. The compound that they become has different properties.

We use a formula to show the elements of a chemical compound. A **chemical formula** is a shorthand way of showing what is in a compound by using symbols and subscripts. The letter symbols let us know what elements are involved and the number subscript tells us how many atoms of each element are involved. No subscript is used if there is only one atom involved. For example, carbon dioxide is made up of one atom of carbon (C) and two atoms of oxygen (O_2), so the formula is CO_2.

TEACHER CERTIFICATION STUDY GUIDE

Sample Test Question and Rationale

11. **Negatively charged particles that circle the nucleus of an atom are called...**
    (Easy)

    A. neutrons.
    B. neutrinos.
    C. electrons.
    D. protons.

Answer: C. electrons.

Neutrons and protons make up the core of an atom. Neutrons have no charge and protons are positively charged. Electrons are the negatively charged particles around the nucleus.

Skill 2.1.2   Atoms, molecules, and chemical bonds

Chemical bonds form when atoms with incomplete valence shells share or completely transfer their valence electrons. There are three types of chemical bonds; covalent bonds, ionic bonds, and hydrogen bonds.

Covalent bonding is the sharing of a pair of valence electrons by two atoms. A simple example of this is two hydrogen atoms. Each hydrogen atom has one valence electron in its outer shell, therefore the two hydrogen atoms come together to share their electrons. Some atoms share two pairs of valence electrons, like two oxygen atoms. This is a double covalent bond.

Electronegativity describes the ability of an atom to attract electrons toward itself. The greater the electronegativity of an atom, the more it pulls the shared electrons towards itself. Electronegativity of the atoms determines whether the bond is polar or nonpolar. In **nonpolar covalent bonds**, the electrons are shared equally, thus the electronegativity of the two atoms is the same. This type of bonding usually occurs between two of the same atoms. A **polar covalent bond** forms when different atoms join, as in hydrogen and oxygen to create water. In this case, oxygen is more electronegative than hydrogen so the oxygen pulls the hydrogen electrons toward itself.

Ionic bonds form when one electron is stripped away from its atom to join another atom. An example of this is sodium chloride (NaCl). A single electron on the outer shell of sodium joins the chloride atom with seven electrons in its outer shell. The sodium now has a +1 charge and the chloride now has a -1 charge. The charges attract each other to form an ionic bond. Ionic compounds are called salts. In a dry salt crystal, the bond is so strong it requires a great deal of strength to break it apart. However, if the salt crystal is placed in water, the bond will dissolve easily as the attraction between the two atoms decreases.

The weakest of the three bonds is the **hydrogen bond**. A hydrogen bond forms when one electronegative atom shares a hydrogen atom with another electronegative atom. An example of a hydrogen bond is a water molecule (H_2O) bonding with an ammonia molecule (NH_3). The H^+ atom of the water molecule attracts the negatively charged nitrogen in a weak bond. Weak hydrogen bonds are beneficial because they can briefly form, the atoms can respond to one another, and then break apart allowing formation of new bonds. Hydrogen bonding plays a very important role in the chemistry of life.

Sample Test Questions and Rationale

12. Which of the following are properties of water?

    I. High specific heat
    II. Strong ionic bonds
    III. Good solvent
    IV. High freezing point
    (Average Rigor)

    A. I, III, IV
    B. II and III
    C. I and II
    D. II, III, IV

Answer: A. I, III, IV

All are properties of water except strong ionic bonds. Water is held together by polar covalent bonds between hydrogen and oxygen atoms.

13. **Potassium chloride is an example of a(n):**
    (Average Rigor)

    A. nonpolar covalent bond.
    B. polar covalent bond.
    C. ionic bond.
    D. hydrogen bond.

Answer: C. ionic bond.

Ionic bonds are formed when one electron is stripped away from its atom to join another atom. Ionic compounds are called salts and potassium chloride is a salt; therefore, potassium chloride is an example of an ionic bond.

Skill 2.1.3 pH and buffers

The pH scale tells how acidic or basic a solution is. An acid is a solution that increases the hydrogen ion concentration of a solution. An example is hydrochloric acid (HCl). A base is a solution that reduces the hydrogen ion concentration. An example is sodium hydroxide (NaOH).

The pH scale ranges from zero to fourteen. Seven is a neutral solution. This is the pH of pure water. A pH between zero and 6.9 is acidic. Stomach acid has a pH of 2.0. A pH between 7.1 and 14 is basic. Common household bleach has a pH of 12.
The internal pH of most living organisms is close to 7. Human blood has a pH of 7.4. Variation from this neutral pH can be harmful to the living organism. Biological fluids resist pH variation due to buffers that minimize the effects of H^+ and OH^- concentrations. A buffer accepts or donates H^+ ions from or to the solution when they are in excess or depleted.

The pH of a substance has a dramatic effect on the environment as well. Acids greatly affect the environment. Acidic precipitation is rain, snow, or fog with a pH less than 5.6. Acidic precipitation is caused by sulfur oxides and nitrogen oxides in the environment that react with water in the air to form acids that fall down to earth as precipitation. A change in pH in the environment can affect the solubility of minerals in the soil, which causes a delay in forest growth.

Sample Test Question and Rationale

14. Sulfur oxides and nitrogen oxides in the environment react with water to cause:
    (Easy)

    A. ammonia.
    B. acidic precipitation.
    C. sulfuric acid.
    D. global warming.

Answer: B. acidic precipitation

Acidic precipitation is rain, snow, or fog with a pH less than 5.6. It is caused by sulfur oxides and nitrogen oxides that react with water in the air to form acids that fall down to Earth as precipitation.

Skill 2.1.4 Biologically important molecules

A compound consists of two or more elements. There are four major chemical compounds found in the cells and bodies of living things. These are carbohydrates, lipids, proteins, and nucleic acids.

Monomers are the simplest unit of structure. **Monomers** combine to form **polymers**, or long chains, making a large variety of molecules. Monomers combine through the process of condensation reactions (also called dehydration synthesis). In this process, one molecule of water is removed between each of the adjoining molecules. In order to break the molecules apart in a polymer, water molecules are added between monomers, thus breaking the bonds between them. This process is called hydrolysis.

Carbohydrates contain a ratio of two hydrogen atoms for each carbon and oxygen $(CH_2O)_n$. Carbohydrates include sugars and starches. They function in the release of energy. **Monosaccharide's** are the simplest sugars and include glucose, fructose, and galactose. They are the major nutrients for cells. In cellular respiration, the cells extract the energy from glucose molecules. **Disaccharides** are made by joining two monosaccharides by condensation to form a glycosidic linkage (covalent bond between two monosaccharides). Maltose is the combination of two glucose molecules, lactose is the combination of glucose and galactose, and sucrose is the combination of glucose and fructose. **Polysaccharides** consist of many monomers joined together and may be structural or provide energy storage for the cell. As energy stores, polysaccharides are hydrolyzed as needed to provide sugar for cells. Examples of polysaccharides include starch, glycogen, cellulose, and chitin.

Starch - major energy storage molecule in plants. It is a polymer consisting of glucose monomers.

Glycogen - major energy storage molecule in animals. It is made up of many glucose molecules.

Cellulose - found in plant cell walls, its function is structural. Many animals lack the enzymes necessary to hydrolyze cellulose, so it simply adds bulk (fiber) to the diet.

Chitin - found in the exoskeleton of arthropods and fungi. Chitin contains an amino sugar (glycoprotein).

Lipids are composed of glycerol (an alcohol) and three fatty acids. Lipids are **hydrophobic** (water fearing) and will not mix with water. There are three important families of lipids; fats, phospholipids and steroids.

Fats consist of glycerol (alcohol) and three fatty acids. Fatty acids are long carbon skeletons. The nonpolar carbon-hydrogen bonds in the tails of fatty acids are highly hydrophobic. Fats are solids at room temperature and come from animal sources (e.g., butter and lard).

Phospholipids are a vital component in cell membranes. In a phospholipid, one or two fatty acids are replaced by a phosphate group linked to a nitrogen group. They consist of a **polar** (charged) head that is hydrophilic (water loving) and a **nonpolar** (uncharged) tail which is hydrophobic. This allows the membrane to orient itself with the polar heads facing the interstitial fluid found outside the cell and the nonpolar tails facing the internal fluid of the cell.

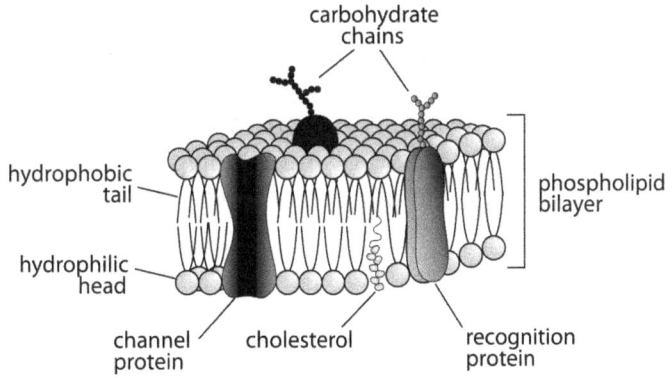

Steroids are insoluble and are composed of a carbon skeleton consisting of four inter-connected rings. An important steroid is cholesterol, which is the precursor from which other steroids are synthesized.

Hormones, including cortisone, testosterone, estrogen, and progesterone, are steroids. Their insolubility keeps them from dissolving in body fluids.

Proteins comprise about fifty percent of the dry weight of animals and bacteria. Proteins function in structure and support (e.g., connective tissue, hair, feathers, and quills), storage of amino acids (e.g., albumin in eggs and casein in milk), transport of substances (e.g. hemoglobin), coordination of body activities (e.g. insulin), signal transduction (e.g. membrane receptor proteins), contraction (e.g., muscles, cilia, and flagella), body defense (e.g. antibodies), and as enzymes to speed up chemical reactions.

All proteins are made of twenty **amino acids**. An amino acid contains an amino group and an acid group. The radical group varies and defines the amino acid. Amino acids form through condensation reactions that remove water. The bond formed between two amino acids is called a peptide bond. Polymers of amino acids are called polypeptide chains. An analogy can be drawn between the twenty amino acids and the alphabet. We can form millions of words using an alphabet of only twenty-six letters. Similarly, organisms can create many different proteins using the twenty amino acids. This results in the formation of many different proteins, whose structure typically defines its function.

There are four levels of protein structure: primary, secondary, tertiary, and quaternary. **Primary structure** is the protein's unique sequence of amino acids. A slight change in primary structure can affect a protein's conformation and its ability to function. **Secondary structure** is the coils and folds of polypeptide chains. The coils and folds are the result of hydrogen bonds along the polypeptide backbone. The secondary structure is either in the form of an alpha helix or a pleated beta sheet. The alpha helix is a coil held together by hydrogen bonds. A beta pleated sheet is the polypeptide chain folding back and forth. The hydrogen bonds between parallel regions hold it together. **Tertiary structure** results from bonds between the side chains of the amino acids. For example, disulfide bridges form when two sulfhydryl groups on the amino acids form a strong covalent bond. **Quaternary structure** is the overall structure of the protein from the aggregation of two or more polypeptide chains. An example of this is hemoglobin. Hemoglobin consists of two kinds of polypeptide chains.

TEACHER CERTIFICATION STUDY GUIDE

Sample Test Question and Rationale

15. What is necessary for diffusion to occur?
    (*Average Rigor*)

    A. carrier proteins
    B. energy
    C. a concentration gradient
    D. a membrane

Answer: C. a concentration gradient

Diffusion is the ability of molecules to move from areas of high concentration to areas of low concentration (a concentration gradient).

Skill 2.1.5   Thermodynamics and free energy

The law of conservation of energy states that energy is neither created nor destroyed. Thus, energy changes form when energy transactions occur in nature. The following are the major forms of energy.

Thermal energy is the total internal energy of objects created by the vibration and movement of atoms and molecules. Heat is the transfer of thermal energy.

Acoustical energy, or sound energy, is the movement of energy through an object in waves. Energy that forces an object to vibrate creates sound.

Radiant energy is the energy of electromagnetic waves. Light, visible and otherwise, is an example of radiant energy.

Electrical energy is the movement of electrical charges in an electromagnetic field. Examples of electrical energy are electricity and lightning.

Chemical energy is the energy stored in the chemical bonds of molecules. For example, the energy derived from gasoline is chemical energy.

Mechanical energy is the potential and kinetic energy of a mechanical system. Rolling balls, car engines, and body parts in motion exemplify mechanical energy.

Nuclear energy is the energy present in the nucleus of atoms. Division, combination, or collision of nuclei release nuclear energy.

Because the total energy in the universe is constant, energy continually transitions between forms. For example, an engine burns gasoline converting the chemical energy of the gasoline into mechanical energy, a plant converts radiant energy of the sun into chemical energy found in glucose, or a battery converts chemical energy into electrical energy.

Interacting objects in the universe constantly exchange and transform energy. Total energy remains the same, but the form of the energy readily changes. Energy often changes from kinetic (motion) to potential (stored) or potential to kinetic. In reality, available energy, energy that is easily utilized, is rarely conserved in energy transformations. Heat energy is an example of relatively "useless" energy often generated during energy transformations. Exothermic reactions release heat and endothermic reactions require heat energy to proceed. For example, the human body is notoriously inefficient in converting chemical energy from food into mechanical energy. The digestion of food is exothermic and produces substantial heat energy.

The three laws of thermodynamics are as follows:

1. The total amount of energy in the universe is constant, energy cannot be created or destroyed, but can merely change form.

    Equation:
    $\Delta E = Q + W$
    Change in energy = (Heat energy entering or leaving) + (work done)

2. In energy transformations, entropy (disorder) increases and useful energy is lost (as heat).

    Equation:
    $\Delta S = \Delta Q / T$
    Change in entropy = (Heat transfer) / (Temperature)

3. As the temperature of a system approaches absolute zero, entropy (disorder) approaches a constant.

Another important concept in the study of thermodynamics is free energy (also called Gibbs free energy). Gibbs free energy is the amount of thermodynamic energy in a system available to do work. The most relevant application of Gibbs free energy is the description of a chemical reaction in an open container at constant temperature and pressure.

The following is the mathematical equation for calculating Gibbs free energy.

$$\Delta G = \Delta H - T\Delta S$$

ΔH is the change in enthalpy in joules, T is the temperature in Kelvin, and ΔS is the change in entropy in joules per Kelvin of the system. Enthalpy, or heat content, is the sum of the thermodynamic energy of a system and the energy the system uses to do work on the environment. The change in Gibbs free energy helps describe the nature of chemical reactions. If ΔG is negative, the reaction is exergonic, proceeds spontaneously, and increases the system's free energy. Conversely, if ΔG is positive, the reaction is endergonic, non-spontaneous, and requires the input of energy to proceed (thus decreasing the system's free energy).

Skill 2.1.6　Cellular bioenergetics

Cellular bioenergetics is the comparison of energy investment and the flow of energy through the cell. In this area of study, we ask if the product is worth the energy investment it requires. In the case of cells, we see that metabolism and reproduction are both worth the energy input because the cell profits in both areas. Photosynthesis results in usable energy, as does cellular respiration. It is important to note that optimal conditions will improve bioenergetics and can sometimes have an effect on the cellular pathway chosen.

For example, let us look at bacteria and plants. Chemosynthetic bacteria are accustomed to living without light, so in lieu of solar energy, they oxidize sulfites or ammonia (chemosynthesis). In some cases, hydrogen atoms are present and the chemical pathway shifts. If hydrogen is available, the bacterium will produce energy from the reaction of hydrogen with carbon dioxide. This energy is enough to fuel the production of biomass, a biological material derived from living organisms that is a renewable energy source. However, if hydrogen is absent from the environment, the bacterium must use a different chemosynthetic pathway. It will utilize energy for chemosynthesis from reactions between O_2 and hydrogen sulfide or ammonia. In this scenario, the chemosynthetic microorganisms are dependent on photosynthesis to occur elsewhere to produce the O_2 they require.

Thus, the environment causes a shift in pathways. The organism utilizes the pathway that is more beneficial (produces the most energy with the least amount of work). Plants are similar in that they can function through either C3 or C4 photosynthesis. The determining factor is their environment. C3 photosynthesis is the typical mechanism of photosynthesis that most plants use. C4 photosynthesis, on the other hand, is an adaptation to arid environmental conditions in which water is used more efficiently. Greater efficiency equals improved cellular bioenergetics.

Skill 2.1.7 Photosynthesis

Photosynthesis is an anabolic process that stores energy in the form of a three carbon sugar. We will use glucose as an example for this section.

Photosynthesis occurs only in organisms that contain chloroplasts (i.e., plants, some bacteria, and some protists). There are a few terms to be familiar with when discussing photosynthesis.

An **autotroph** (self-feeder) is an organism that makes its own food from the energy of the sun or other elements. Autotrophs include:

1. **photoautotrophs** - make food from light and carbon dioxide, releasing oxygen that can be used for respiration.

2. **chemoautotrophs** - oxidize sulfur and ammonia. Some bacteria are chemoautotrophs.

Heterotrophs (other feeder) are organisms that must eat other living things to obtain energy. Another term for heterotrophs is **consumers**. All animals are heterotrophs. **Decomposers** break down once living things. Bacteria and fungi are examples of decomposers. **Scavengers** eat dead things. Examples of scavengers are bacteria, fungi, and some animals.

The **chloroplast** is the site of photosynthesis in a plant cell. Similar to mitochondria in a eukaryotic cell, chloroplasts contain an increased surface area of membrane called the thylakoid membrane. The thylakoid membrane contains pigments (chlorophyll) that are capable of capturing light energy. Between the membranous stacks of thylakoids there is a fluid called stroma.
Photosynthesis reverses the electron flow. Water is split by the chloroplast into hydrogen and oxygen. Oxygen is released as a waste product as carbon dioxide is reduced to sugar (glucose). This requires the input of energy, which comes from the sun.

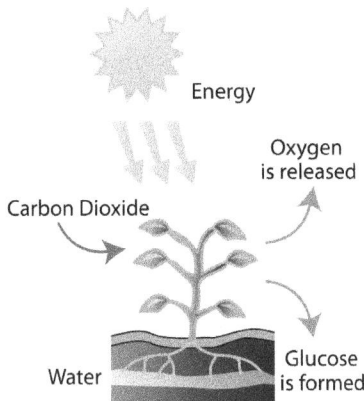

Photosynthesis occurs in two stages, the light reactions and the Calvin cycle (dark reactions). The conversion of solar energy to chemical energy occurs during light reactions. In light reactions, chlorophyll absorbs light and uses the energy to split water, releasing oxygen as a waste product. The conversion of light energy to chemical energy is stored in the form of NADPH and ATP. Both NADPH and ATP are then used in the Calvin cycle to produce sugar.

The second stage of photosynthesis is the **Calvin cycle**. Carbon dioxide in the air is incorporated into organic molecules already in the chloroplast. The NADPH produced in the light reaction is used as reducing power for the reduction of the carbon to carbohydrate. ATP from the light reaction is also needed to convert carbon dioxide to carbohydrate (sugar).

The two stages of photosynthesis are summarized below.

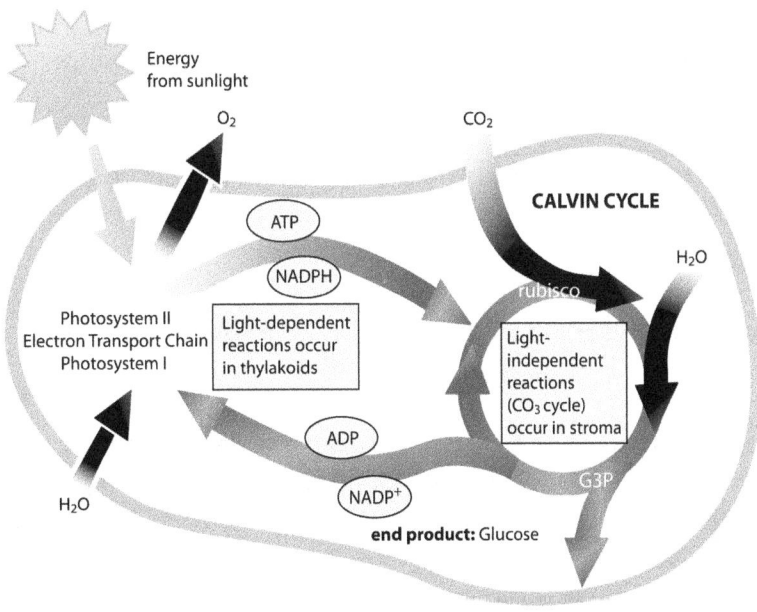

The process of photosynthesis is made possible by the sun. Visible light ranges in wavelengths of 750 nanometers (red light) to 380 nanometers (violet light). As the wavelength decreases, the amount of available energy increases. Light is carried as photons, which are fixed quantities of energy. Light is reflected (what we see), transmitted, or absorbed (what the plant uses). The plant's pigments capture light of specific wavelengths. Remember that the reflected light is what we see as color. Plant pigments include:

>Chlorophyll *a* - reflects green/blue light; absorbs red light
>Chlorophyll *b* - reflects yellow/green light; absorbs red light
>Carotenoids - reflects yellow/orange; absorbs violet/blue light

The pigments absorb photons. The energy from the light excites electrons in the chlorophyll that jump to orbitals with more potential energy and reach an "excited" or unstable state.

The formula for photosynthesis is:

$$CO_2 + H_2O + \text{energy (from sunlight)} \rightarrow C_6H_{12}O_6 \text{ (glucose)} + O_2$$

The high energy electrons are trapped by primary electron acceptors, which are located on the thylakoid membrane. These electron acceptors and the pigments form reaction centers called photosystems that are capable of capturing light energy. Photosystems contain a reaction-center chlorophyll that releases an electron to the primary electron acceptor. This transfer is the first step of the light reactions. There are two photosystems, named accordingly by their date of discovery, not their order of occurrence.

>**Photosystem I** is composed of a pair of chlorophyll *a* molecules. Photosystem I is also called P700 because it absorbs light of 700 nanometers. Photosystem I makes ATP whose energy is needed to build glucose.

>**Photosystem II** is also called P680 because it absorbs light of 680 nanometers. Photosystem II produces ATP + $NADPH_2$ and the waste gas oxygen.

Both photosystems are bound to the **thylakoid membrane**, close to electron acceptors.

The production of ATP is termed **photophosphorylation** due to the use of light. Photosystem I uses cyclic photophosphorylation because the pathway occurs in a cycle. It can also use noncyclical photophosphorylation, which starts with light and ends with glucose. Photosystem II uses noncyclical photophosphorylation only.

Below is a diagram of the relationship between photosynthesis and cellular respiration, which will be covered in the following section.

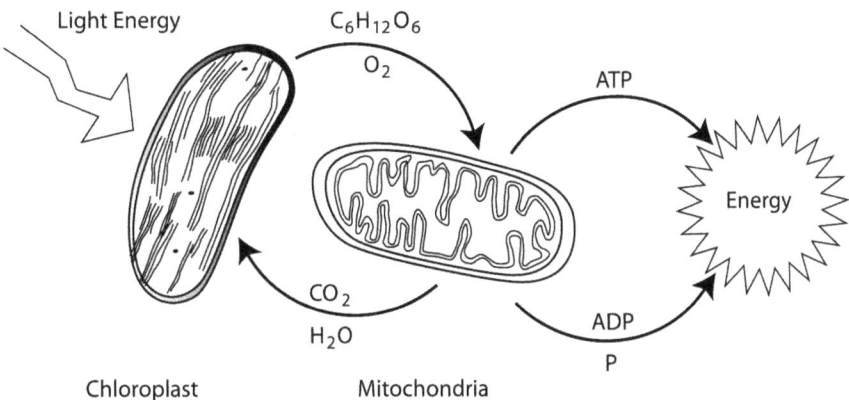

Skill 2.1.8 Respiration

Cellular respiration is the metabolic pathway in which food (e.g. glucose) is broken down to produce energy in the form of ATP. Both plants and animals utilize respiration to create energy for metabolism. In respiration, energy is released by the transfer of electrons in a process known as an **oxidation-reduction (redox)** reaction. The oxidation phase of this reaction is the loss of an electron and the reduction phase is the gain of an electron. Redox reactions are important for all stages of respiration.

Glycolysis is the first step in respiration. It occurs in the cytoplasm of the cell and does not require oxygen. Each of the ten stages of glycolysis is catalyzed by a specific enzyme. The following is a summary of these stages.

In the first stage the reactant is glucose. For energy to be released from glucose, it must be converted to a reactive compound. This conversion occurs through the phosphorylation of a molecule of glucose by the use of two molecules of ATP. This is an investment of energy by the cell. The 6-carbon product, called fructose -1,6- bisphosphate, breaks into two 3-carbon molecules of sugar. A phosphate group is added to each sugar molecule and hydrogen atoms are removed. Hydrogen is picked up by NAD^+ (a vitamin). Since there are two sugar molecules, two molecules of NADH are formed. The reduction (addition of hydrogen) of NAD allows the potential for energy transfer.

As the phosphate bonds are broken, ATP is produced. Two ATP molecules are generated as each original 3-carbon sugar molecule is converted to pyruvic acid (pyruvate). A total of four ATP molecules are made in the four stages. Since two molecules of ATP are needed to start the reaction in stage 1, there is a net gain of two ATP molecules at the end of glycolysis. This accounts for only two percent of the total energy in a molecule of glucose.

Beginning with pyruvate, which was the end product of glycolysis, the following steps occur before entering the **Krebs cycle**.

1. Pyruvic acid is changed to acetyl-CoA (coenzyme A). This is a 3-carbon pyruvic acid molecule, which has lost one molecule of carbon dioxide (CO_2) to become a 2-carbon acetyl group. Pyruvic acid loses a hydrogen to NAD^+, which is reduced to NADH.
2. Acetyl CoA enters the Krebs cycle. For each molecule of glucose it started with, two molecules of Acetyl CoA enter the Krebs cycle (one for each molecule of pyruvic acid formed in glycolysis).

The **Krebs cycle** (also known as the citric acid cycle), occurs in four major steps. First, the 2-carbon acetyl CoA combines with a 4-carbon molecule to form a 6-carbon molecule of citric acid. Next, two carbons are lost as carbon dioxide (CO_2) and a 4-carbon molecule is formed, which is available to join with CoA to form citric acid again. Since we started with two molecules of CoA, two turns of the Krebs cycle are necessary to process the original molecule of glucose. In the third step, eight hydrogen atoms are released and picked up by FAD and NAD (vitamins and electron carriers).

Lastly, for each molecule of CoA (remember there were two to start with) you get:

> 3 molecules of NADH x 2 cycles
> 1 molecule of $FADH_2$ x 2 cycles
> 1 molecule of ATP x 2 cycles

Therefore, this completes the breakdown of glucose. At this point, a total of four molecules of ATP have been made, two from glycolysis and one from each of the two turns of the Krebs cycle. Six molecules of carbon dioxide have been released, two prior to entering the Krebs cycle and two for each of the two turns of the Krebs cycle. Twelve carrier molecules have been made, ten NADH and two $FADH_2$. These carrier molecules will carry electrons to the electron transport chain.

ATP is made by substrate level phosphorylation in the Krebs cycle. Notice that the Krebs cycle in itself does not produce many ATP molecules. Instead, it functions mostly in the transfer of electrons that are subsequently used in the electron transport chain to generate large numbers of ATP molecules. In the **Electron Transport Chain,** NADH transfers electrons from glycolysis and the Krebs cycle to the first molecule in the chain of molecules embedded in the inner membrane of the mitochondrion.

Most of the molecules in the electron transport chain are proteins. Nonprotein molecules are also part of the chain and are essential for the catalytic functions of certain enzymes. The electron transport chain does not make ATP directly. Instead, it breaks up a large free energy drop into a more manageable one. The chain uses electrons to pump H^+ ions across the mitochondrial membrane. The H^+ gradient is used to form ATP synthesis in a process called **chemiosmosis** (oxidative phosphorylation). ATP synthetase and energy, generated by the movement of hydrogen ions coming from NADH and $FADH_2$, builds ATP from ADP on the inner membrane of the mitochondria. Each NADH yields three molecules of ATP (10 x 3) and each $FADH_2$ yields two molecules of ATP (2 x 2). Thus, the electron transport chain and oxidative phosphorylation produces 34 ATP and the net gain from the whole process of respiration is 36 molecules of ATP.

Process	# ATP produced (+)	# ATP consumed (-)	Net # ATP
Glycolysis	4	2	+2
Acetyl CoA	0	2	-2
Krebs cycle	1 per cycle (2 cycles)	0	+2
Electron transport chain	34	0	+34
Total			+36

Aerobic versus anaerobic respiration

Glycolysis generates ATP with oxygen (aerobic) or without oxygen (anaerobic). We have already discussed aerobic respiration. Anaerobic respiration occurs through fermentation. In the process of fermentation, ATP is generated by substrate level phosphorylation if enough NAD^+ is present to accept electrons during oxidation. In anaerobic respiration, NAD^+ is regenerated by transferring electrons to pyruvate. There are two common types of fermentation.

In **alcoholic fermentation**, pyruvate is converted to ethanol in two steps. In the first step, carbon dioxide is released from the pyruvate. In the second step, ethanol is produced when acetaldehyde is reduced by NADH. This results in the regeneration of NAD^+ for glycolysis. Alcohol fermentation is carried out by yeast and some bacteria.

In **lactic acid fermentation**, pyruvate is reduced by NADH forming lactate as a waste product. Animal cells and some bacteria that do not use oxygen utilize lactic acid fermentation to make ATP. Lactic acid forms when pyruvic acid accepts hydrogen from NADH. A buildup of lactic acid is what causes muscle soreness following exercise.

Energy remains stored in lactic acid or alcohol until it is needed. This is not an efficient type of respiration. When oxygen is present, aerobic respiration occurs after glycolysis.

Both aerobic and anaerobic pathways oxidize glucose to pyruvate through the process of glycolysis and both pathways employ NAD^+ as an oxidizing agent. A substantial difference between the two pathways is that in fermentation, an organic molecule such as pyruvate or acetaldehyde is the final electron acceptor. In respiration, the final electron acceptor is oxygen. Another key difference is that respiration yields much more energy from a sugar molecule than fermentation does. Respiration can produce up to 18 times more ATP than fermentation.

Sample Test Questions and Rationale

16. The loss of an electron is _____ and the gain of an electron is _____.
    (Rigorous)

    A. oxidation, reduction
    B. reduction, oxidation
    C. glycolysis, photosynthesis
    D. photosynthesis, glycolysis

Answer: A. oxidation, reduction

Oxidation-reduction reactions are also known as redox reactions. In respiration, energy is released by the transfer of electrons in redox reactions. The oxidation phase of this reaction involve the loss of an electron and the reduction phase involves the gain of an electron.

17. During the Krebs cycle, 8 carrier molecules are formed. What are they?
    (Rigorous)

    A. 3 NADH, 3 FADH, 2 ATP
    B. 6 NADH and 2 ATP
    C. 4 $FADH_2$ and 4 ATP
    D. 6 NADH and 2 $FADH_2$

Answer: D. 6 NADH and 2 $FADH_2$

For each molecule of CoA that enters the Kreb's cycle, you get 3 NADH and 1 $FADH_2$. There are 2 molecules of CoA so the total yield is 6 NADH and 2 $FADH_2$ during the Krebs cycle.

18. The product of anaerobic respiration in animals is:
    (Average Rigor)

    A. carbon dioxide.
    B. lactic acid.
    C. pyruvate.
    D. ethyl alcohol.

Answer: B. lactic acid.

In anaerobic lactic acid fermentation, pyruvate is reduced by NADH to form lactic acid. This is the anaerobic process in animals. Alcoholic fermentation is an anaerobic process in yeast and some bacteria yielding ethyl alcohol. Carbon dioxide and pyruvate are products of aerobic respiration

Skill 2.1.9   Enzymes

Enzymes act as biological catalysts that speed up reactions. Enzymes are the most diverse type of protein. They are not used up in a reaction and are recyclable. Each enzyme is specific for a single reaction. Enzymes act on a substrate. The substrate is the material to be broken down or put back together. Most enzymes end in the suffix -ase (lipase, amylase). The prefix is the substrate being acted on (lipids, sugars).

$$\text{Substrate} \xrightarrow{\text{Enzyme}} \text{Product}$$

The active site is the region of the enzyme that binds to the substrate. There are two theories for how the active site functions. The **lock and key theory** states that the shape of the enzyme is specific because it fits into the substrate like a key fits into a lock. In this theory, the enzyme holds molecules close together so reactions can easily occur. The **Induced fit theory** states that an enzyme can stretch and bend to fit the substrate. This is the most accepted theory.

Many factors can affect enzyme activity. Temperature and pH are two such factors. The temperature can affect the rate of reaction of an enzyme. The optimal pH for enzymes is between 6 and 8, with a few enzymes whose optimal pH falls outside of this range.

Cofactors aid in the enzyme's function. Cofactors may be inorganic or organic. Organic cofactors are known as coenzymes. Vitamins are examples of coenzymes. Some chemicals can inhibit an enzyme's function. **Competitive inhibitors** block the substrate from entering the active site of the enzyme to reduce productivity. **Noncompetitive inhibitors** bind to a location on the enzyme that is different from the active site and interrupts substrate binding. In most cases, noncompetitive inhibitors alter the shape of the enzyme. An **allosteric enzyme** can exist in two shapes; they are active in one form and inactive in the other. Overactive enzymes may cause metabolic diseases.

Sample Test Question and Rationale

19. ATP is known to bind to phosphofructokinase-1 (an enzyme involved in glycolysis). This results in a change in the shape of the enzyme that causes the rate of ATP production to fall. Which answer best describes this phenomenon?
    (Rigorous)

    A. binding of a coenzyme
    B. an allosteric change in the enzyme
    C. competitive inhibition
    D. uncompetitive inhibition

Answer: B. an allosteric change in the enzyme

The binding of ATP to phosphofructokinase-1 causes an allosteric change (a change in shape) of the enzyme. The binding of ATP can be considered non-competitive inhibition.

Competency 2.2 Cell Structure and function

Skill 2.2.1 Membranes, organelles, and subcellular components of cells

The cell is the basic unit of all living things. There are three types of cells: archaea, prokaryotic, and eukaryotic. Archaea have some similarities with prokaryotes, but are as distantly related to prokaryotes as prokaryotes are to eukaryotes.

ARCHAEA

There are three kinds of organisms with archaea cells: **methanogens**, obligate anaerobes that produce methane, **halobacteria**, which can live only in concentrated brine solutions, and **thermoacidophiles**, which can live only in acidic hot springs.

PROKARYOTES

Prokaryotes consist only of bacteria and cyanobacteria (formerly known as blue-green algae). The diagram below depicts the classification of prokaryotes.

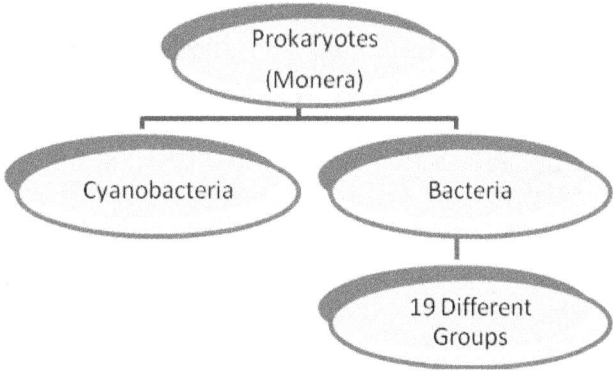

Bacterial cells have no defined nucleus or nuclear membrane. The DNA, RNA, and ribosomes float freely within the cell. The cytoplasm has a single chromosome condensed to form a **nucleoid**. Prokaryotes have a thick cell wall made up of amino sugars (glycoproteins) that provides protection, gives the cell shape, and keeps the cell from bursting. The antibiotic penicillin targets the **cell wall** of bacteria. Penicillin works by disrupting the cell wall, thus killing the cell. The cell wall surrounds the **cell membrane** (plasma membrane). The cell membrane consists of a lipid bilayer that controls the passage of molecules in and out of the cell. Some prokaryotes have a capsule, made of polysaccharides, surrounding the cell wall for extra protection from higher organisms.

Many bacterial cells have appendages used for movement called **flagella**. Some cells also have **pili**, which are a protein strand used for attachment. Pili are also used for sexual conjugation (where bacterial cells exchange DNA).

Prokaryotes are the most numerous and widespread organisms on earth. Bacteria were most likely the first cells and date back in the fossil record to 3.5 billion years ago. Their ability to adapt to the environment allows them to thrive in a wide variety of habitats.

EUKARYOTES

Eukaryotic cells are found in protists, fungi, plants, and animals. Most eukaryotic cells are larger than prokaryotic cells. They contain many organelles, which are membrane-bound areas for specific functions. Their cytoplasm contains a cytoskeleton, which provides a protein framework for the cell. The cytoplasm also supports the organelles and contains the ions and molecules necessary for cell function. The cytoplasm is contained by the plasma membrane. The plasma membrane allows molecules to pass in and out of the cell. The membrane can bud inward to engulf outside material in a process called endocytosis. Exocytosis is a secretory mechanism, the reverse of endocytosis.

The most significant differentiation between prokaryotes and eukaryotes is that eukaryotes have a **nucleus**. The nucleus is the "brain" of the cell that contains all of the cell's genetic information. The chromosomes consist of chromatin, which are complexes of DNA and proteins. The chromosomes are tightly coiled to conserve space while providing a large surface area. The nucleus is the site of transcription of the DNA into RNA. The **nucleolus** is where ribosomes are made. There is at least one of these dark-staining bodies inside the nucleus of most eukaryotes. The nuclear envelope consists of two membranes separated by a narrow space. The envelope contains many pores that let RNA out of the nucleus.

Ribosomes are the site for protein synthesis. Ribosomes may be free floating in the cytoplasm or attached to the endoplasmic reticulum. There may be up to a half a million ribosomes in a cell, depending on how much protein the cell makes.

The **endoplasmic reticulum** (ER) is folded and has a large surface area. It is the "roadway" of the cell and allows for transport of materials through and out of the cell. There are two types of ER: smooth and rough. Smooth endoplasmic reticulum does not contain ribosomes on the surface and is the site of lipid synthesis. Rough endoplasmic reticulum has ribosomes on its surface and aids in the synthesis of proteins that are membrane-bound or destined for secretion.

Many of the products made in the ER proceed to the Golgi apparatus. The **Golgi apparatus** functions to sort, modify, and package molecules that are made in other parts of the cell (like the ER). These molecules are sent either out of the cell or to other organelles within the cell.

Lysosomes are found mainly in animal cells. These contain digestive enzymes that break down food, unnecessary substances, viruses, damaged cell components, and, eventually, the cell itself. It is believed that lysosomes play a role in the aging process.

Mitochondria are large organelles that are the site of cellular respiration, the process of ATP production that supplies energy to the cell. Muscle cells have many mitochondria because they use a great deal of energy. Mitochondria have their own DNA, RNA, and ribosomes and are capable of reproducing by binary fission if there is a great demand for additional energy. Mitochondria have two membranes: a smooth outer membrane and a folded inner membrane. The folds inside the mitochondria are called cristae. They provide a large surface area for cellular respiration to occur.

Plastids are found only in photosynthetic organisms. They are similar to mitochondria in that they both have a double membrane structure. They also have their own DNA, RNA, and ribosomes and can reproduce if the need for the increased capture of sunlight becomes necessary. There are several types of plastids. **Chloroplasts** are the site of photosynthesis. The stroma is the chloroplast's inner membrane space. The stroma encloses sacs called thylakoids that contain the photosynthetic pigment chlorophyll. The chlorophyll traps sunlight inside the thylakoid to generate ATP, which is used in the stroma to produce carbohydrates and other products. The **chromoplasts** make and store yellow and orange pigments. They provide color to leaves, flowers, and fruits. The **amyloplasts** store starch and are used as a food reserve. They are abundant in roots like potatoes.

The Endosymbiotic Theory states that mitochondria and chloroplasts were once free living and possibly evolved from prokaryotic cells. At some point in our evolutionary history, they entered the eukaryotic cell and maintained a symbiotic relationship with the cell. The fact that they both have their own DNA, RNA, ribosomes, and are capable of reproduction supports this theory.

Found only in plant cells, the **cell wall** is composed of cellulose and fibers. It is thick enough for support and protection, yet porous enough to allow water and dissolved substances to enter. **Vacuoles** are found mostly in plant cells. They hold stored food and pigments. Their large size allows them to fill with water in order to provide turgor pressure. Lack of turgor pressure causes a plant to wilt. The **cytoskeleton**, found in both animal and plant cells, is composed of protein filaments attached to the plasma membrane and organelles. The cytoskeleton provides a framework for the cell and aids in cell movement. Three types of fibers make up the cytoskeleton:

1. **Microtubules** – The largest of the three fibers, they make up cilia and flagella for locomotion. Some examples are sperm cells, cilia that line the fallopian tubes, and tracheal cilia. Centrioles are also composed of microtubules. They aid in cell division to form the spindle fibers that pull the cell apart into two new cells. Centrioles are not found in the cells of higher plants.

2. **Intermediate filaments** – Intermediate in size, they are smaller than microtubules, but larger than microfilaments. They help the cell keep its shape.

3. **Microfilaments** – Smallest of the three fibers, they are made of actin and small amounts of myosin (like in muscle tissue). They function in cell movement like cytoplasmic streaming, endocytosis, and amoeboid movement. This structure pinches the two cells apart after cell division, forming two new cells.

The following is a diagram of a generalized animal cell.

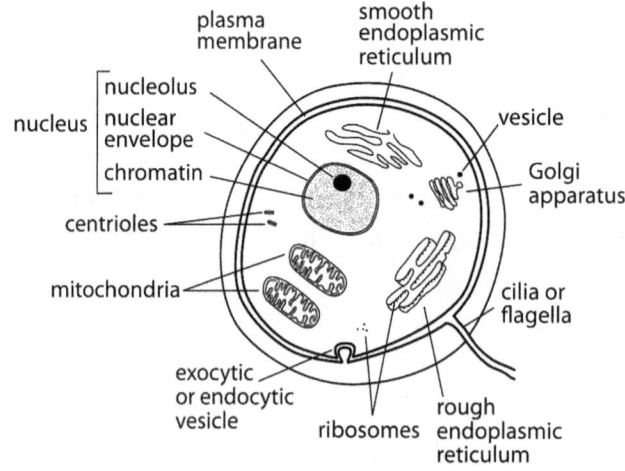

TEACHER CERTIFICATION STUDY GUIDE

Sample Test Questions and Rationale

20. The shape of a cell depends on its...
    (Average Rigor)

    A. function.
    B. structure.
    C. age.
    D. size.

Answer: A. function.

In most living organisms, cellular structure is based on function.

21. Which type of cell would contain the most mitochondria?
    (Average Rigor)

    A. muscle cell
    B. nerve cell
    C. epithelial cell
    D. blood cell

Answer: A. muscle cell

Mitochondria are the site of cellular respiration where ATP is produced. Muscle cells have the most mitochondria because they use a great deal of energy.

22. Which of the follow is not true of both chloroplasts and mitochondria?
    (Easy)

    A. the inner membrane is the primary site for it's activity
    B. converts energy from one form to another
    C. uses an electron transport chain
    D. is an important part of the carbon cycle

Answer: A. the inner membrane is the primary site for it's activity

In mitochondria the electron transport chain is present in the inner membrane, however in chloroplasts it is present in the thylakoid membranes.

BIOLOGY

23. Which part of the cell is responsible for lipid synthesis?
    (Rigorous)

    A. Golgi Apparatus
    B. Rough Endoplasmic Reticulum
    C. Smooth Endoplasmic Reticulum
    D. Lysosome

Answer: C. Smooth Endoplasmic Reticulum

The rough endoplasmic reticulum and the golgi apparatus are both involved in the production of proteins (synthesis and packaging, respectively). Lysosomes contain digestive enzymes. Only the smooth endoplasmic reticulum is directly responsible for lipid production.

24. According to the fluid-mosaic model of the cell membrane, membranes are composed of
    (Rigorous)

    A. a phospholipid bilayer with proteins embedded in the layers.
    B. one layer of phospholipids with cholesterol embedded in the layer
    C. two layers of protein with lipids embedded in the layers.
    D. DNA and fluid proteins.

Answer: A. a phospholipid bilayer with proteins embedded in the layers.

Cell membranes are composed of a phospholipid bilayer in which hydrophilic heads face outward and hydrophobic tails are sandwiched between the hydrophilic layers. The membrane contains proteins embedded in the layer (integral proteins) and proteins on the surface (peripheral proteins).

25. A type of molecule not found in the membrane of an animal cell is
    (Rigorous)

    A. phospholipid.
    B. protein.
    C. cellulose.
    D. cholesterol.

Answer: C. cellulose.

Phospholipids, protein, and cholesterol are all found in animal cells. Cellulose, however, is only found in plant cells.

26. Which of the following is not considered evidence of Endosymbiotic Theory?
    (Rigorous)

    A. The presence of genetic material in mitochondria and plastids.
    B. The presence of ribosomes within mitochondria and plastids.
    C. The presence of a double layered membrane in mitochondria and plastids.
    D. The ability of mitochondria and plastids to reproduce.

Answer: C. The presence of a double layered membrane in mitochondria and plastids.

A double layered membrane is not unique to mitochondria and plastids, the nucleus is also double layered. All three other characteristics are not present in any other organelle, and are evidence that mitochondria and plastids may once have been separate organisms.

Skill 2.2.2 Cell cycle, cytokinesis, mitosis, and meiosis

The purpose of cell division is to provide growth and repair in body (somatic) cells and to replenish or create sex cells for reproduction. There are two forms of cell division: mitosis and meiosis. **Mitosis** is the division of somatic cells and **meiosis** is the division of sex cells (eggs and sperm).

Mitosis is divided into two parts: the **mitotic (M) phase** and **interphase**. In the mitotic phase, mitosis and cytokinesis divide the nucleus and cytoplasm, respectively. The mitotic phase is the shortest phase of the cell cycle. Interphase is the stage where the cell grows and copies the chromosomes in preparation for the mitotic phase. Interphase occurs in three stages of growth: the G_1 (growth) period, when the cell grows and metabolizes, the **S** (synthesis) period, when the cell makes new DNA, and the G_2 (growth) period, when the cell makes new proteins and organelles in preparation for cell division.

The mitotic phase is a continuum of change, although we divide it into five distinct stages: prophase, prometaphase, metaphase, anaphase, and telophase.

During **prophase**, the cell proceeds through the following steps continuously, without stopping. First, the chromatin condenses to become visible chromosomes. Next, the nucleolus disappears and the nuclear membrane breaks apart. Then, mitotic spindles, composed of microtubules, form that will eventually pull the chromosomes apart. Finally, the cytoskeleton breaks down and the centrioles push the spindles to the poles or opposite ends of the cell.

During **prometaphase**, the nuclear membrane fragments and allows the spindle microtubules to interact with the chromosomes. Kinetochore fibers attach to the chromosomes at the centromere region. **Metaphase** begins when the centrosomes are at opposite ends of the cell. The centromeres of all the chromosomes are aligned with one another.

During **anaphase**, the centromeres split in half and homologous chromosomes separate. The chromosomes are pulled to the poles of the cell, with identical sets at either end. The last stage of mitosis is **telophase**. Here, two nuclei form with a full set of DNA that is identical to the parent cell. The nucleoli become visible and the nuclear membrane reassembles. A cell plate is seen in plant cells and a cleavage furrow forms in animal cells. The cell pinches into two cells. Finally, cytokinesis, or division of the cytoplasm and organelles, occurs. Below is a diagram of mitosis.

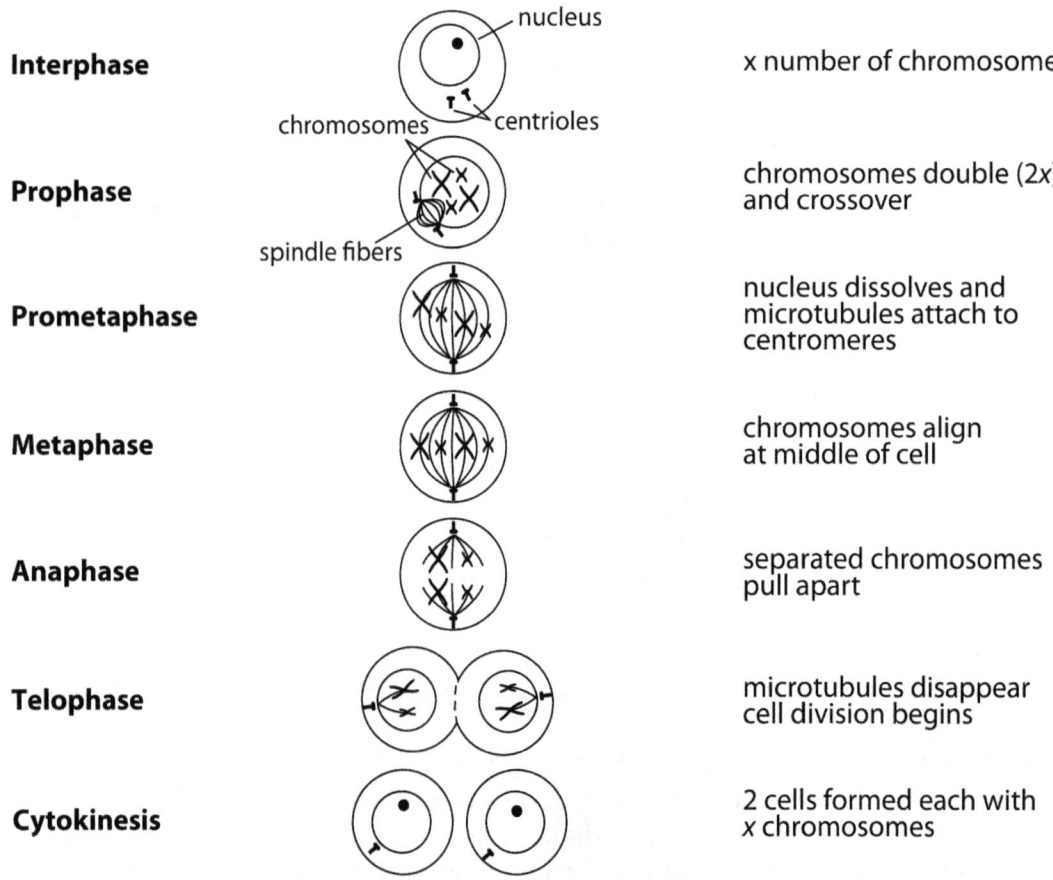

Meiosis is similar to mitosis, but there are two consecutive cell divisions, meiosis I and meiosis II, in order to reduce the chromosome number by one-half. This way, when the sperm and egg join during fertilization, the diploid number is reached.

Similar to mitosis, meiosis is preceded by an interphase during which the chromosome replicates. The steps of meiosis are as follows:

1. **Prophase I** – The replicated chromosomes condense and pair with homologues in a process called synapsis. This forms a tetrad. Crossing over, the exchange of genetic material between homologues to further increase diversity, occurs during prophase I.

2. **Metaphase I** – The homologous pairs attach to spindle fibers after lining up in the middle of the cell.

3. **Anaphase I** – The sister chromatids remain joined and move to the poles of the cell.

4. **Telophase I** – The homologous chromosome pairs continue to separate. Each pole now has a haploid chromosome set. Telophase I occurs simultaneously with cytokinesis. In animal cells, a cleavage furrow forms and, in plant cells, a cell plate appears.

5. **Prophase II** – A spindle apparatus forms and the chromosomes condense.

6. **Metaphase II** – Sister chromatids line up in center of cell. The centromeres divide and the sister chromatids begin to separate.

7. **Anaphase II** – The separated chromosomes move to opposite ends of the cell.

8. **Telophase II** – Cytokinesis occurs, resulting in four haploid daughter cells.

Meiosis

Interphase	x number of chromosomes
Prophase	chromosomes double (2x) and crossover
Prometaphase	nucleus dissolves and microtubules attach to centromeres
Metaphase I	chromosomes align at middle of cell
Anaphase I	separated chromosomes pull apart
Telophase I	microtubules disappear cell division begins
Prophase II	2 cells formed each with x chromosomes
Metaphase II	microtubules attach to centromeres
Anaphase II	chromosomes pull apart
Telophase II	microtubules disappear cell division begins
Cytokinesis	4 cells form each with half the number of original chromosomes (½ x)

BIOLOGY

TEACHER CERTIFICATION STUDY GUIDE

Sample Test Questions and Rationale

27. Identify this stage of mitosis.

(Average Rigor)

A. anaphase
B. metaphase
C. telophase
D. prophase

Answer: B. metaphase

During metaphase, the centromeres are at opposite ends of the cell. During this phase the chromosomes are aligned with one another in the middle of the cell.

28. **Which statement regarding mitosis is correct?**
(Easy)

A. diploid cells produce haploid cells for sexual reproduction
B. sperm and egg cells are produced
C. diploid cells produce diploid cells for growth and repair
D. it allows for greater genetic diversity

Answer: C. diploid cells produce diploid cells for growth and repair

The purpose of mitotic cell division is to provide growth and repair in body (somatic) cells. The cells begin as diploid and produce diploid cells.

29. This stage of mitosis includes cytokinesis or division of the cytoplasm and its organelles.
    (Average Rigor)

    A. anaphase
    B. interphase
    C. prophase
    D. telophase

Answer: D. telophase

The last stage of the mitosis is telophase. Here, the two nuclei form with a full set of DNA each. The cell is pinched in half into two cells and cytokinesis, or the division of the cytoplasm and organelles, occurs.

30. Replication of chromosomes occurs during which phase of the cell cycle?
    (Average Rigor)

    A. prophase
    B. interphase
    C. metaphase
    D. anaphase

Answer: B. interphase

Interphase is the stage where the cell grows and copies the chromosomes in preparation for the mitotic phase.

31. Which process(es) result(s) in a haploid chromosome number?
    (Easy)

    A. both meiosis and mitosis
    B. mitosis
    C. meiosis
    D. replication and division

Answer: C. meiosis

In meiosis, there are two consecutive cell divisions resulting in the reduction of chromosome number by half (diploid to haploid).

TEACHER CERTIFICATION STUDY GUIDE

32. **Crossing over, which increases genetic diversity, occurs during which stage(s)?**
    (Rigorous)

    A. telophase II in meiosis
    B. metaphase in mitosis
    C. interphase in both mitosis and meiosis
    D. prophase I in meiosis

Answer: D. prophase I in meiosis

During prophase I of meiosis, the replicated chromosomes condense and pair with their homologues in a process called synapsis. Crossing over, the exchange of genetic material between homologues, occurs during prophase I.

Competency 2.3 Molecular basis of heredity

Skill 2.3.1 Structure and function of nucleic acids

Nucleic acids consist of DNA (deoxyribonucleic acid) and RNA (ribonucleic acid). Nucleic acids contain the code for the amino acid sequence of proteins and the instructions for replication. The monomer of nucleic acids is a nucleotide. A nucleotide consists of a 5-carbon sugar (deoxyribose in DNA, ribose in RNA), a phosphate group, and a nitrogenous base. The base sequence is the code or the instructions. There are five bases: adenine, thymine, cytosine, guanine, and uracil. Uracil is found only in RNA and replaces thymine. The following provides a summary of nucleic acid structure:

	SUGAR	PHOSPHATE	BASES
DNA	Deoxyribose	Present	adenine, **thymine**, cytosine, guanine
RNA	Ribose	Present	adenine, **uracil**, cytosine, guanine

Due to the molecular structure, adenine will always pair with thymine in DNA or uracil in RNA. Cytosine always pairs with guanine in both DNA and RNA.

BIOLOGY

This allows for the symmetry of the DNA molecule seen below.

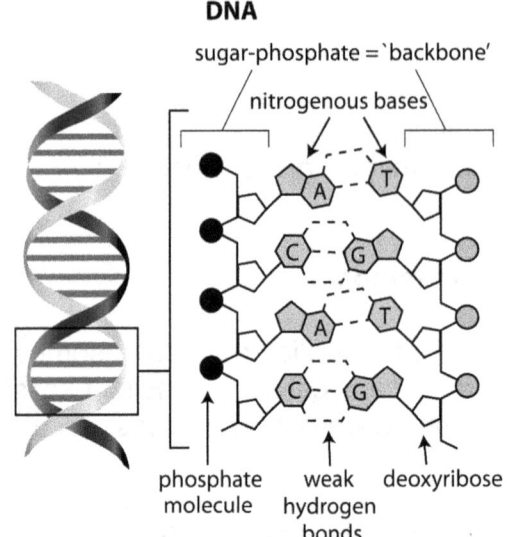

Adenine and thymine (or uracil in RNA) are linked by two hydrogen bonds and cytosine and guanine are linked by three hydrogen bonds. Guanine and cytosine are harder to break apart than thymine (uracil) and adenine because of the greater number of bonds between the bases. The double-stranded DNA molecule forms a double helix, or twisted ladder, shape.

Skill 2.3.2   DNA replication

DNA replicates semi conservatively meaning the two original DNA strands are conserved and serve as a template for the new strand.

In DNA replication, the first step is to separate the two strands. As they separate, they unwind the supercoils to reduce tension. An enzyme called **helicase** unwinds the DNA as the replication fork proceeds and **topoisomerases** relieve the tension by nicking one strand and relaxing the supercoil.

Once the strands have separated, they must be stabilized. Single-strand binding proteins (SSBs) bind to the single strands until the DNA is replicated.

An RNA polymerase called primase adds ribonucleotides to the DNA template to initiate DNA synthesis. This short RNA-DNA hybrid is called a **primer**. Once the DNA is single stranded, **DNA polymerases** add nucleotides in the 5' ® 3' direction.

As DNA synthesis proceeds along the replication fork, it becomes obvious that replication is semi-discontinuous; meaning one strand is synthesized in the direction the replication fork is moving and the other is synthesized in the opposite direction. The continuously synthesized strand is the **leading strand** and the discontinuously synthesized strand is the **lagging strand**. As the replication fork proceeds, new primer is added to the lagging strand and it is synthesized discontinuously in small fragments called **Okazaki fragments**.

The remaining RNA primers must be removed and replaced with deoxyribonucleotides. DNA polymerase has 5' ⑧ 3' polymerase activity and has 3' ⑧ 5' exonuclease activity. This enzyme binds to the nick between the Okazaki fragment and the RNA primer. It removes the primer and adds deoxyribonucleotides in the 5' ⑧ 3' direction. The nick remains until **DNA ligase** seals it, producing the final product, a double-stranded segment of DNA.

Once the double-stranded segment is replicated, there is a proofreading system carried out by DNA replication enzymes. In eukaryotes, DNA polymerases have 3' ⑧ 5' exonuclease activity—they move backwards and remove nucleotides when the enzyme recognizes an error, then add the correct nucleotide in the 5' ⑧ 3' direction. In bacteria, DNA polymerase III is the main polymerase that elongates DNA during replication and has exonuclease proofreading ability.

DNA Replication

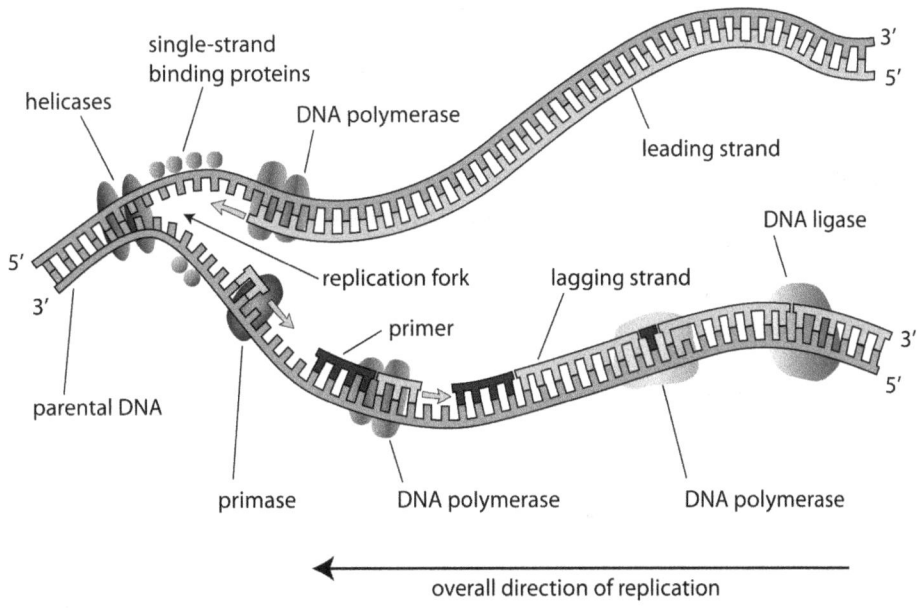

Chromosomal replication in bacteria is similar to eukaryotic DNA replication.

A **plasmid** is a small ring of DNA that carries accessory genes separate from those of the bacterial chromosome. Most plasmids in Gram-negative bacteria undergo bidirectional replication, although some replicate unidirectional because of their small size. Plasmids in Gram-positive bacteria replicate by the rolling circle mechanism.

Some plasmids can transfer themselves (and therefore their genetic information) through a process called conjugation. Conjugation requires cell-to-cell contact. The sex pilus of the donor cell attaches to the recipient cell. Once contact has been established, the transfer of DNA occurs by the rolling circle mechanism.

Sample Test Questions and Rationale

33. **DNA synthesis results in a strand that is synthesized continuously. This is the...**
    (Average Rigor)

    A. lagging strand.
    B. leading strand.
    C. template strand.
    D. complementary strand.

Answer: B. leading strand.

As DNA synthesis proceeds along the replication fork, one strand is replicated continuously (the leading strand) and the other strand is replicated discontinuously (the lagging strand).

34. **Segments of DNA can be transferred from one organism to another through the use of which of the following?**
    (Average Rigor)

    A. bacterial plasmids
    B. viruses
    C. chromosomes from frogs
    D. plant DNA

Answer: A. bacterial plasmids

Plasmids can transfer themselves (and therefore their genetic information) through a process called conjugation. This requires cell-to-cell contact.

Skill 2.3.3 Protein synthesis

Proteins are synthesized through the process of translation. Three major classes of RNA are needed to carry out these processes: messenger RNA (mRNA), ribosomal RNA (rRNA), and transfer RNA (tRNA). **Messenger RNA** contains information for translation. **Ribosomal RNA** is a structural component of the ribosome and **transfer RNA** carries amino acids to the ribosome for protein synthesis.

Transcription is similar in prokaryotes and eukaryotes. During transcription, the DNA molecule is copied into an RNA molecule (mRNA). Transcription occurs through the steps of initiation, elongation, and termination. Transcription also occurs for rRNA and tRNA, but the focus here is on mRNA.

Initiation begins at the promoter of the double-stranded DNA molecule. The promoter is a specific region of DNA that directs the **RNA polymerase** to bind to the DNA. The double-stranded DNA opens up and RNA polymerase begins transcription in the 5' ⑧ 3' direction by pairing ribonucleotides to the deoxyribonucleotides as follows to get a complementary mRNA segment:

Deoxyribonucleotide		Ribonucleotide
A	⑧	U
G	⑧	C

Elongation is the synthesis of the mRNA strand in the 5' ⑧ 3' direction. The new mRNA rapidly separates from the DNA template and the complementary DNA strands pair together.

Termination of transcription occurs at the end of a gene. Cleavage occurs at specific sites on the mRNA. This process is aided by termination factors.

In eukaryotes, mRNA goes through **posttranscriptional processing** before going on to translation.

There are three basic steps of processing:

1. **5' capping** – The addition of a base with a methyl attached to it that protects the 5' end from degradation and serves as the site where ribosomes bind to the mRNA for translation.

2. **3' polyadenylation** – The addition of 100-300 adenines to the free 3' end of mRNA resulting in a poly-A-tail.

3. **Intron removal**- The removal of non-coding introns and the splicing together of coding exons form the mature mRNA.

TEACHER CERTIFICATION STUDY GUIDE

Translation is the process in which the mRNA sequence becomes a polypeptide. The mRNA sequence determines the amino acid sequence of a protein by following a pattern called the genetic code. The **genetic code** consists of 64 triplet nucleotide combinations called **codons**. Three codons are termination codons and the remaining 61 code for amino acids. There are 20 amino acids encoded by different mRNA codons. Amino acids are the building blocks of protein. They are attached together by peptide bonds to form a polypeptide chain.

Ribosomes are the site of translation. They contain rRNA and many proteins. Translation occurs in three steps: initiation, elongation, and termination. Initiation occurs when the methylated tRNA binds to the ribosome to form a complex. This complex then binds to the 5' cap of the mRNA. In elongation, tRNAs carry the amino acid to the ribosome and place it in order according to the mRNA sequence. tRNA is very specific – it only accepts one of the 20 amino acids that correspond to the anticodon. The anticodon is complementary to the codon. For example, using the codon sequence below:

the mRNA reads   A U G / G A G / C A U / G C U
the anticodons are   U A C / C U C / G U A / C G A

Termination occurs when the ribosome reaches any one of the three stop codons: UAA, UAG, or UGA. The newly formed polypeptide then undergoes posttranslational modification to alter or remove portions of the polypeptide.

Sample Test Questions and Rationale

35. **Which of the following is not a form of posttranscriptional processing?**
    (Rigorous)

    A. 5' capping
    B. intron splicing
    C. polypeptide splicing
    D. 3' polyadenylation

Answer: C. polypeptide splicing

The removal of segments of polypeptides is a posttranslational process. The other three are methods of posttranscriptional processing.

36. This carries amino acids to the ribosome in protein synthesis.
    (Average Rigor)

    A. messenger RNA
    B. ribosomal RNA
    C. transfer RNA
    D. DNA

Answer: C. transfer RNA

The tRNA molecule is specific for a particular amino acid. tRNA has an anticodon sequence that is complementary to the codon. This specifies where the tRNA places the amino acid in protein synthesis.

37. A DNA molecule has the sequence ACTATG. What is the anticodon of this molecule?
    (Rigorous)

    A. UGAUAC
    B. ACUAUG
    C. TGATAC
    D. ACTATG

Answer: B. ACUAUG

The DNA is first transcribed into mRNA. Here, the DNA has the sequence ACTATG; therefore, the complementary mRNA sequence is UGAUAC (remember, in RNA, T is U). This mRNA sequence is the codon. The anticodon is the complement to the codon. The anticodon sequence will be ACUAUG (remember, the anticodon is tRNA, so U is present instead of T).

Skill 2.3.4 Gene regulation

In bacterial cells, the *lac* operon is a good example of the control of gene expression. The *lac* operon contains the genes that code for the enzymes used to convert lactose into fuel (glucose and galactose). The *lac* operon contains three genes, *lac Z*, *lac Y*, and *lac A*. *Lac Z* codes for an enzyme that converts lactose into glucose and galactose. *Lac Y* codes for an enzyme that causes lactose to enter the cell. *Lac A* codes for an enzyme that acetylates lactose.

The *lac* operon also contains a promoter and an operator that is the "off and on" switch for the operon. A protein called the repressor switches the operon off when it binds to the operator. When lactose is absent, the repressor is active and the operon is turned off. The operon is turned on again when allolactose (formed from lactose) inactivates the repressor by binding to it.

Skill 2.3.5 Mutation and transposable elements

Inheritable changes in DNA are called mutations. **Mutations** may be errors in replication or a spontaneous rearrangement of one or more nucleotide segments by factors such as radioactivity, drugs, or chemicals. The severity of the change is not as critical as where the change occurs. DNA contains large segments of non-coding areas called introns. The important coding areas are called exons. If an error occurs in an intron, it has no effect. If the error occurs in an exon, it may range from having no effect to being lethal depending on the severity of the mistake. Mutations may occur on somatic or sex cells. Usually mutations in sex cells are more dangerous since they contain the blueprint for the developing offspring. However, mutations are not always bad. They are the basis of evolution and if they create a favorable variation that enhances the organism's survival, they are beneficial. However, mutations may also lead to abnormalities, birth defects, and even death. There are several types of mutations.

A **point mutation** is a mutation involving a single nucleotide or a few adjacent nucleotides. Let us suppose a normal sequence was as follows:

Normal	A B C D E F	
Duplication	A B **C C** D E F	one nucleotide is repeated
Inversion	A **E D C B** F	a segment of the sequence is flipped
Deletion	A B C E F	a nucleotide is left out (D is lost)
Insertion	A B C **R S** D E F	nucleotides are inserted or translocated
Breakage	A B C	a piece is lost (DEF is lost)

Deletion and insertion mutations that shift the reading frame are called **frame shift mutations**.

A **silent mutation** alters the nucleotide sequence but does not change the amino acid sequence, therefore it does not alter the protein. A **missense mutation** results in an alteration in the amino acid sequence. A mutation's effect on protein function depends on which amino acids are involved and how many are involved. The structure of a protein usually determines its function. A mutation that does not alter the structure will probably have little or no effect on the protein's function. However, a mutation that does alter the structure of a protein and severely affects protein activity is called a **loss-of-function mutation**. Sickle-cell anemia and cystic fibrosis are examples of loss-of-function mutations.

TEACHER CERTIFICATION STUDY GUIDE

Skill 2.3.6   Viruses

Microbiology includes the study of monera, protists, and viruses. Although **viruses** are not classified as living things, they greatly affect other living things by disrupting cell activity. Viruses are obligate parasites because they rely on the host for their own reproduction. Viruses are composed of a protein coat and a nucleic acid, either DNA or RNA. A bacteriophage is a virus that infects a bacterium. Animal viruses are classified by the type of nucleic acid, presence of RNA replicase, and presence of a protein coat.

There are two types of viral reproductive cycles:

1. **Lytic cycle** - The virus enters the host cell and makes copies of its nucleic acid and protein coat, and then reassembles. Afterward, it lyses or breaks out of the host cell and infects other nearby cells, repeating the process.
2. **Lysogenic cycle** - The virus may remain dormant within the cell until some factor activates and stimulates it to break out of the cell. Herpes is an example of a lysogenic virus.

Sample Test Question and Rationale

38. Viruses are made of...
    (Easy)

    A. a protein coat surrounding nucleic acid.
    B. DNA, RNA, and a cell wall.
    C. nucleic acid surrounding a protein coat.
    D. protein surrounded by DNA.

Answer: A. a protein coat surrounding nucleic acid.

Viruses are composed of a protein coat and nucleic acid; either RNA or DNA.

Skill 2.3.7   Molecular basis of genetic diseases, e.g., cancer, sickle-cell anemia, hemophilia

Sickle-cell anemia is characterized by weakness, heart failure, joint and muscular impairment, fatigue, abdominal pain and dysfunction, impaired mental function, and eventual death. The mutation that causes this genetic disorder is a point mutation in the sixth amino acid of hemoglobin. A normal hemoglobin molecule has glutamic acid as the sixth amino acid and the sickle-cell hemoglobin has valine at the sixth position. This mutation causes the chemical properties of hemoglobin to change. The hemoglobin of a sickle-cell person has a lower affinity for oxygen, causing red blood cells to have a sickle shape.

BIOLOGY

The sickle shape of the red blood cell causes the formation of clogs because the cells have difficulty passing through capillaries.

Cystic fibrosis is the most common genetic disorder of people with European ancestry. This disorder affects the exocrine system. A fibrous cyst forms on the pancreas that blocks the pancreatic ducts. This causes sweat glands to release high levels of salt. A thick mucus, secreted from mucous glands, accumulates in the lungs. The accumulation of mucous causes bacterial infections and possibly death. Cystic fibrosis cannot be cured, but can be treated for a short time. Most children with cystic fibrosis die before adulthood. Scientists have identified a mutation in the protein that transports chloride ions across cell membranes in patients with cystic fibrosis. The majority of the mutant alleles have a deletion of the three nucleotides coding for phenylalanine at position 508. Other people with the disorder have mutant alleles caused by substitution, deletion, and frameshift mutations.

Hemophilia is an inheritable genetic disorder characterized by a body's inability to clot. Individuals with hemophilia have lower levels of clotting factors present in their blood. When a blood vessel is damaged, a clot will not form, causing excessive bleeding. This is possible in both external (e.g. skin) and internal (e.g., muscles, joints, or brain) injuries. Hemophilia is a heritable trait because it is passed from mother to child (most commonly male offspring) on the maternal X chromosome. The paternal Y chromosome does not have a gene for this trait. Nevertheless, the paternal X chromosome has the ability to block the trait (dominant) or, by not blocking it, to passively enable it. Hemophilia is an X-linked trait because it is always associated with a deficiency on the X chromosome. Women have two X chromosomes. As long as one chromosome is active, she will be unaffected, although she will be a carrier and is capable of passing on the recessive defective gene. Because a male has only one X chromosome, and the Y has no gene for this trait, he is far more likely to be affected by his deficient gene, thus hemophilia is more common in men. Hemophilia is split into three categories: severe, moderate, and mild as determined by the levels of clotting factors present in the individual's blood.

Unrestricted cell cycle and cancer

The restriction point in the cell cycle occurs late in the G_1 phase of the cell cycle. This is when the decision for the cell to divide is made. If all the internal and external cell systems are working properly, the cell proceeds to replicate. Cells may also decide not to proceed past the restriction point. This nondividing cell state is called the G_0 phase. Many specialized cells remain in this state.

The density of cells also regulates cell division. Density-dependent inhibition describes when cells crowd one another and consume all available nutrients, thereby halting cell division. Cancer cells do not respond to density-dependent inhibition. They divide excessively and invade other tissues. As long as there are nutrients, cancer cells are "immortal."

Sample Test Question and Rationale

39. **Cancer cells divide extensively and invade other tissues. This continuous cell division is due to:**
    (Rigorous)

    A. density dependent inhibition.
    B. density independent inhibition.
    C. chromosome replication.
    D. growth factors.

Answer: B. density independent inhibition.

Density dependent inhibition is when the cells crowd one another and consume all the nutrients, thereby halting cell division. Cancer cells, however, are density independent; meaning they can divide continuously as long as nutrients are present.

Skill 2.3.8   Recombinant DNA and genetic engineering

Genetic engineering has made enormous contributions to medicine and has enabled significant enhancements in DNA technology.

The use of DNA probes and the polymerase chain reaction (PCR) has enabled scientists to identify and detect elusive pathogens. Diagnosis of genetic disease is now possible before the onset of symptoms.

Genetic engineering has allowed for the treatment of some genetic disorders. **Gene therapy** is the introduction of a normal allele into somatic cells to replace a defective allele. The medical field has had success in treating patients with a single enzyme deficiency. Gene therapy has allowed doctors and scientists to introduce a normal allele providing the missing enzyme.

Insulin and mammalian growth hormones have been produced in bacteria by gene-splicing techniques. Insulin treatment helps control diabetes for millions of people who suffer from the disease. The insulin produced in genetically engineered bacteria is chemically identical to that made in the pancreas. Human growth hormone (HGH) has been genetically engineered for the treatment of dwarfism caused by insufficient amounts of HGH. HGH is being further researched for treatment of broken bones and severe burns.

Biotechnology has advanced the techniques used to create vaccines. Genetic engineering allows for the modification of a pathogen in order to attenuate it for vaccine use. In fact, vaccines created by a pathogen attenuated by gene-splicing may be safer than those that use traditional mutants.

In its simplest form, genetic engineering requires enzymes to cut DNA, a vector, and a host organism in which to place the recombinant DNA. A **restriction enzyme** is a bacterial enzyme that cuts foreign DNA in specific locations. The restriction fragment that results can be inserted into a bacterial plasmid (**vector**). Other vectors that may be used include viruses and bacteriophages. The splicing of restriction fragments into a plasmid results in a recombinant plasmid. This recombinant plasmid can then be placed in a host cell, usually a bacterial cell, for replication.

The use of recombinant DNA provides a means to transplant genes among species. This opens the door for cloning specific genes of interest. Hybridization can be used to find a gene of interest. A probe is a molecule complementary to the sequence of a gene of interest. The probe, once annealed to the gene, can be detected by labeling with a radioactive isotope or a fluorescent tag.

Gel electrophoresis is another method for analyzing DNA. Electrophoresis separates DNA or protein by size or electrical charge. The DNA runs towards the positive charge and the DNA fragments separate by size. The gel is treated with a DNA-binding dye that fluoresces under ultraviolet light. A picture of the gel can be taken and used for analysis.

One of the most widely used genetic engineering techniques is the **polymerase chain reaction (PCR)**. PCR is a technique in which a piece of DNA can be amplified into billions of copies within a few hours. This process requires a primer to specify the segment to be copied, and an enzyme (usually taq polymerase) to amplify the DNA. PCR has allowed scientists to perform multiple procedures on small amounts of DNA.

TEACHER CERTIFICATION STUDY GUIDE

Sample Test Questions and Rational

40. **Which of the following is not a useful application of genetic engineering?**
    (Rigorous)

    A. The creation of safer viral vaccines.
    B. The creation of bacteria that produce hormones for medical use.
    C. The creation of bacteria to breakdown toxic waste.
    D. The creation of organisms that are successfully being used as sources for alternative fuels.

Answer: D. The creation of organisms that are successfully being used as a source alternative fuels.

Although there is a push to genetically engineer organisms that will either create alternative fuels or be used as an alternative fuel source, this field is in its infancy. There are multiple successful examples for each of the other posssible answers.

41. **Which of the following is a way that cDNA cloning has not been used?**
    (Rigorous)

    A. to provide evidence for taxonomic organization
    B. to study the mutations that lead to diseases such as hemophilia
    C. to determine the structure of a protein
    D. to understand methods of gene regulation

Answer: C. to determine the structure of a protein

Although cDNA cloning can be used to determine the amino acid sequence of a protein many other steps determine the final protein structure. For example, the folding of the protein, addition of other protein subunits, and/or modification by other proteins.

42. **A genetic engineering advancement in the medical field is...**
    (Easy)

    A. gene therapy.
    B. pesticides.
    C. degradation of harmful chemicals.
    D. antibiotics.

Answer: A. gene therapy.

Gene therapy is the introduction of a normal allele imto somatic cells in order to replace a defective gene. The medical field has had success in treating patients with a single enzyme deficiency disease. Gene therapy has allowed doctors and scientists to introduce a normal allele that provides the missing enzyme.

Skill 2.3.9 Genome mapping of humans and other organisms

The goal of the human genome project is to map and sequence the three billion nucleotides in the human genome, and to identify all of the genes on it. The project was launched in 1986 and an outline of the genome was finished in 2000 through international collaboration. In May 2006, the sequence of the last chromosome was published. While the map and sequencing are complete, scientists are still studying the functions of all the genes and their regulation. Humans have successfully decoded the genome of other mammals as well.

TEACHER CERTIFICATION STUDY GUIDE

DOMAIN 3.0 CLASSICAL GENETICS AND EVOLUTION

Competency 3.1 Classical Genetics

Skill 3.1.1 Mendelian inheritance and probability

Gregor Mendel is recognized as the father of genetics. His work in the late 1800's is the basis of our knowledge of genetics. Although unaware of the presence of DNA or genes, Mendel realized there were factors (now known as **genes**) that were transferred from parents to their offspring. Mendel worked with pea plants and fertilized the plants himself, keeping track of subsequent generations which led to the Mendelian laws of genetics. Mendel found that two "factors" governed each trait, one from each parent. Traits or characteristics came in several forms, known as **alleles**. For example, the trait of flower color had white alleles (*pp*) and purple alleles (*PP*). Mendel formulated two laws: the law of segregation and the law of independent assortment.

The **law of segregation** states that only one of the two possible alleles from each parent is passed on to the offspring. If the two alleles differ, then one is fully expressed in the organism's appearance (the dominant allele) and the other has no noticeable effect on appearance (the recessive allele). The two alleles for each trait segregate into different gametes. A Punnet square can be used to show the law of segregation. In a Punnet square, one parent's genes are put at the top of the box and the other parent's on the side. Genes combine in the squares just like numbers are added in addition tables. This Punnet square shows the result of the cross of two F_1 hybrids.

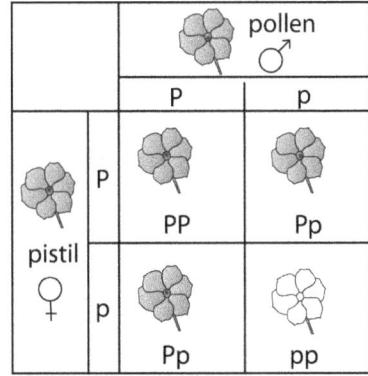
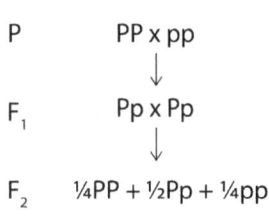

This cross results in a 1:2:1 ratio of F_2 offspring. Here, the *P* is the dominant allele and the *p* is the recessive allele. The F_1 cross produces three offspring expressing the dominant allele (one *PP* and two *Pp*) and one offspring expressing the recessive allele (*pp*). Some other important terms to know:

Homozygous – having a pair of identical alleles. For example, *PP* and *pp* are homozygous pairs.

BIOLOGY 63

Heterozygous – having two different alleles. For example, *Pp* is a heterozygous pair.

Phenotype – the organism's physical appearance.

Genotype – the organism's genetic makeup. For example, *PP* and *Pp* have the same phenotype (purple in color), but different genotypes.

The **law of independent assortment** states that alleles assort independently of each other. The law of segregation applies for monohybrid crosses (only one character, in this case flower color, is experimented with). In a dihybrid cross, two characters are explored. Two of the seven characters Mendel studied were seed shape and color. Yellow is the dominant seed color (*Y*) and green is the recessive color (*y*). The dominant seed shape is round (*R*) and the recessive shape is wrinkled (*r*). A cross between a plant with yellow round seeds (*YYRR*) and a plant with green wrinkled seeds (*yyrr*) produces an F_1 generation with the genotype *YyRr*. Independent assortment of the *YyRr* genotype yields four possible outcomes (*YR*, *Yr*, *yR*, and *yr*). Crossing the F_1 generation would result in the production of F_2 offspring with a 9:3:3:1 phenotypic ratio.

		pollen ♂			
		YR	Yr	yR	yr
pistil ♀	YR	YYRR	YYRr	YyRR	YyRr
	Yr	YYRr	YYrr	YyRr	Yyrr
	yR	YyRR	YyRr	yyRR	yyRr
	yr	YyRr	Yyrr	yyRr	yyrr

P: YYRR x yyrr
F_1: YyRr
F_2:
1 - YYRR
2 - YYRr
2 - YyRR
4 - YyRr } 9 yellow round

1 - yyRR
2 - yyRr } 3 green round

1 - YYrr
2 - Yyrr } 3 yellow wrinkled

1 - yyrr } 1 green wrinkled

Based on Mendelian genetics, the more complex hereditary pattern of **dominance** was discovered. In Mendel's law of segregation, the F_1 generation has either purple or white flowers. This is an example of **complete dominance**. **Incomplete dominance** is when the F_1 generation results in an appearance somewhere between the two parents. For example, red flowers are crossed with white flowers, resulting in an F_1 generation with pink flowers. The red and white traits are still carried by the F_1 generation, resulting in an F_2 generation with a phenotypic ratio of 1:2:1. In **codominance**, the genes may form new phenotypes. ABO blood grouping is an example of codominance.

A and B are of equal strength and O is recessive. Therefore, type A blood may have the genotypes of AA or AO, type B blood may have the genotypes of BB or BO, type AB blood has the genotype A and B, and type O blood has two recessive O genes.

A family pedigree is a collection of a family's history for a particular trait. As you work your way through the pedigree of interest, Mendelian inheritance theories are applied. In tracing a trait, the generations are mapped in a pedigree chart, similar to a family tree but with the alleles present. In a case where both parents have a particular trait and one of two children also express this trait, then the trait is due to a dominant allele. In contrast, if both parents do not express a trait and one of their children does, that trait is due to a recessive allele.

Sample Test Questions and Rationale

43. The Law of Segregation defined by Mendel states…
(Average Rigor)

  A. when sex cells form, the two alleles that determine a trait will end up on different gametes.
  B. only one of two alleles is expressed in a heterozygous organism.
  C. the allele expressed is the dominant allele.
  D. alleles of one trait do not affect the inheritance of alleles on another chromosome.

Answer: A. when sex cells form, the two alleles that determine a trait will end up on different gametes.

The law of segregation states that the two alleles of each trait segregate to different gametes.

TEACHER CERTIFICATION STUDY GUIDE

44. A child with type O blood has a father with type A blood and a mother with type B blood. The genotypes of the parents, respectively, would be which of the following?
    (Average Rigor)

    A. AA and BO
    B. AO and BO
    C. AA and BB
    D. AO and OO

Answer: B. AO and BO

Type O blood has 2 recessive O genes. A child receives one allele from each parent; therefore, each parent in this example must have an O allele. The father has type A blood with a genotype of AO and the mother has type B blood with a genotype of BO.

Skill 3.1.2 Non-Mendelian inheritance

Non-Mendelian inheritance is a general term describing any pattern of genetic inheritance that does not conform to Mendel's laws or does not rely on a single chromosomal gene. Examples of non-Mendelian inheritance include complex traits, environmental influence, organelle DNA, transmission bias, and epigenetics.

Multiple genes determine the expression of many complex traits. For example, disorders arising from a defect in a single gene are rare compared to complex disorders like cancer, heart disease, and diabetes. The inheritance of such complex disorders does not follow Mendelian rules because they involve more than one gene.

While chromosomal DNA carries the majority of an organism's genetic material, organelles, including mitochondria and chloroplasts, also have DNA containing genes. Organelle genes have their own patterns of inheritance that do not conform to Mendelian rules. Such patterns of inheritance are often called maternal because offspring receive all of their organelle DNA from the mother.

Transmission bias describes a situation in which the alleles of the parent organisms are not equally represented in their offspring. Transmission bias often results from the failure of alleles to segregate properly during cell division. Mendelian genetics assumes equal representation of parent alleles in the offspring generation.

BIOLOGY

Epigenetic inheritance involves changes not involving DNA sequence. For example, the addition of methyl groups (methylation) to DNA molecules can influence the expression of genes and override Mendelian patterns of inheritance.

Finally, genetic linkage, discussed in detail in the next section, is often considered a form of non-Mendelian inheritance because closely linked chromosomal genes tend to assort together, not separately. Linkage, however, is not entirely non-Mendelian because classical genetics can generally explain and predict the patterns of inheritance of linked traits.

Skill 3.1.3   Linkage

Genetic linkage is the inheritance of two or more traits together. In general, the transmission of a particular allele is independent of the alleles passed on for other traits. Independent inheritance results from the random sorting of chromosomes during meiosis. Genes found on the same chromosome, however, often remain together during meiosis. Thus, these linked genes have a greater probability of appearing together in offspring.

The phenomenon known as crossing over prevents complete linkage of genes on the same chromosome. During meiosis, paired chromosomes exchange genetic material creating new combinations of DNA. Crossing over is more likely to disrupt linkage when genes are far apart on a chromosome. Greater distance between genes increases the probability that crossing over will occur between the gene loci.

Skill 3.1.4   Human genetic disorders

The same techniques of pedigree analysis apply when tracing inherited disorders. Thousands of genetic disorders result from the inheritance of a recessive trait. These disorders range from non-lethal traits (such as albinism) to life threatening traits (such as cystic fibrosis).

Most people with recessive disorders are born to parents with normal phenotypes. The mating of heterozygous parents results in an offspring genotypic ratio of 1:2:1; thus 1 out of 4 offspring will express the recessive trait. The heterozygous parents are called carriers because they do not express the trait phenotypically but can pass the trait on to their offspring.

Lethal dominant alleles are much less common than lethal recessives. This is because lethal dominant alleles are not masked in heterozygotes. Mutations in a gene of the sperm or egg can result in a lethal dominant allele, usually killing the developing offspring.

Sex linked traits - The Y chromosome found only in males (XY) carries very little genetic information relative to the X chromosome found in females (XX). Since men do not have a second X chromosome to mask recessive genes, recessive traits are expressed more often in men. Women must have recessive genes on both X chromosomes to phenotypically show the trait. Examples of sex-linked traits include hemophilia and color-blindness.

Sex influenced traits - Traits are influenced by sex hormones. Male pattern baldness is an example of a sex-influenced trait. Testosterone influences the expression of the gene, thus, men are more susceptible to hair loss.

Nondisjunction - During meiosis, chromosomes fail to separate properly. One sex cell may get both chromosomes and another may not get any. Depending on the chromosomes involved, this may or may not be serious. Offspring end up with either a missing chromosome or an extra chromosome. An example of nondisjunction is Down Syndrome, where three copies of chromosome 21 are present.

Chromosome Theory - Introduced by Walter Sutton in the early 1900's. In the late 1800's, the processes of mitosis and meiosis were defined. Sutton understood how these processes confirmed Mendel's "factors". The chromosome theory basically states that genes are located on chromosomes that undergo independent assortment and segregation.

Screening for genetic disorders

Some genetic disorders can be prevented. Parents can be screened for genetic disorders before the child is conceived or in the early stages of pregnancy. Genetic counselors determine the risk of producing offspring that may express a genetic disorder. The counselor reviews the family's pedigree and determines the frequency of recessive alleles. While genetic counseling is helpful for future parents, it does not affirmatively ascertain whether a child will present a disorder.

There are some genetic disorders that can be discovered in a heterozygous parent. For example, sickle-cell anemia and cystic fibrosis alleles can be discovered in carriers by genetic testing. If the parents are carriers but decide to have children anyway, fetal testing is available during the pregnancy. There are a few techniques available to determine if a developing fetus will have certain genetic disorders.

Amniocentesis is a procedure in which a needle is inserted into the uterus to extract some of the amniotic fluid surrounding the fetus. Some disorders can be detected by chemicals found in amniotic fluid. Other disorders can be detected by karyotyping cells cultured from the fluid.

TEACHER CERTIFICATION STUDY GUIDE

A physician removes some of the fetal tissue from the placenta in a technique called **chorionic villus sampling (CVS)**. The cells are then karyotyped as they are in amniocentesis. The advantage of CVS is that the cells can be karyotyped immediately, unlike in amniocentesis which takes several weeks to culture.

Unlike amniocentesis and CVS, **ultrasounds** are a non-invasive technique employed to detect genetic disorders. However, ultrasounds are limited in that they can only detect physical abnormalities of the fetus.

Newborn screening is now routinely performed in the United States at birth. One disease that is screened for is Phenylketonuria. Phenylketonuria (PKU) is a recessively inherited disorder that does not allow children to metabolize the amino acid phenylalanine. This amino acid and its by-product accumulate in the blood to toxic levels, resulting in mental retardation. This can be prevented by screening at birth for this defect and treating it with a special diet.

Sample Test Question and Rationale

45. Amniocentesis is:
    (Average Rigor)

    A. a non-invasive technique for detecting genetic disorders.
    B. a bacterial infection.
    C. extraction of amniotic fluid.
    D. removal of fetal tissue.

Answer: C. extraction of amniotic fluid.

Amniocentesis is a procedure in which a needle is inserted into the uterus to extract some of the amniotic fluid surrounding the fetus. Some genetic disorders can be detected by chemicals in the fluid.

Skill 3.1.5   Interaction between heredity and the environment

The environment can have an impact on an individual's phenotype. For example, a person living at a higher altitude will have a different amount of red and white blood cells than a person living at sea level.

In some cases, a particular trait is advantageous to the organism in a particular environment. Sickle-cell disease causes a low oxygen level in the blood, which results in red blood cells having a sickle shape. About one in every ten African-Americans have the sickle-cell trait. Heterozygous carriers are usually healthy compared to homozygous individuals who suffer severe detrimental effects. In tropical African environments, heterozygote people are more resistant to malaria than people who do not carry any copies of the sickle-cell gene.

Sample Test Questions and Rationale

46. A woman has Pearson Syndrome, a disease caused by a mutation in mitochondrial DNA. Which of the following individuals would you expect to see the disease in? *(Rigorous)*

   I   Her Daughter
   II Her Son
   III Her Daughter's son
   IV Her Son's daughter

   A. I, III
   B. I, II, III
   C. II, IV
   D. I, II, III, IV

Answer: B. I, II, III

Since mitochondrial DNA is passed through the maternal line, both of her children would be affected and the trait would continue to pass from her daughter to all of her children. Her son's children would recieve their mitochondrial DNA from their mother.

47. Which is not a possible effect of polyploidy? *(Rigorous)*

   A. More robust members of an animal species.
   B. The creation of cross species offspring.
   C. The creation of a new species.
   D. Cells that produce higher levels of desired proteins.

Answer: A. More robust members of an animal species.

While polyploidy often creates new plant species thereby yeilding more robust crops, it is likely to create nonviable animal offspring.

48.

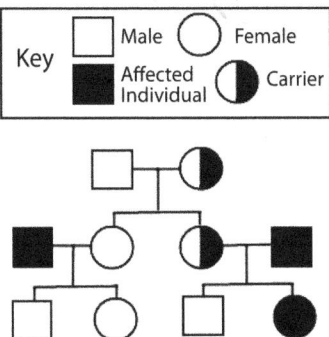

Based on the pedigree chart above, what term best describes the nature of the trait being mapped?
(Rigorous)

    A. autosomal recessive
    B. sex-linked
    C. incomplete dominance
    D. co-dominance

Answer: B. sex-linked

This chart would be a good example of color blindness, a sex-linked trait. If the trait had been autosomal recessive the last generation would all be carriers with the exception of the affected individual. In the case of traits that are incompletely dominant or co-dominant the tree would require additional notation.

Competency 3.2 Evolution

Skill 3.2.1 Evidence

Fossils are the key to understanding biological history. They are the preserved remnants left by an organism that lived in the past. Scientists have established the geological time scale to determine the age of a fossil. The geological time scale is broken down into four eras: Precambrian, Paleozoic, Mesozoic, and Cenozoic. The eras are further broken down into periods that represent a distinct age in the Earth's history. Scientists use rock layers called strata to date fossils. The older layers of rock are at the bottom. This allows scientists to correlate rock layers with the era they date back to. Radiometric dating is a more precise method of dating fossils. Rocks and fossils contain isotopes of elements accumulated over time. The isotope's half-life is used to date older fossils by determining the amount of isotope remaining and comparing it to the isotypes half-life.

TEACHER CERTIFICATION STUDY GUIDE

Dating fossils is helpful in the construction of evolutionary trees. Scientists can arrange the succession of animals based on their fossil record. The fossils of an animal's ancestors can be dated and placed on its evolutionary tree. For example, the branched evolution of horses shows that the modern horse's ancestors were larger, had a reduced number of toes, and had teeth modified for grazing.

Sample Test Question and Rationale

49. Which aspect of science does not support evolution?
    (Average Rigor)

    A. comparative anatomy
    B. organic chemistry
    C. comparison of DNA among organisms
    D. analogous structures

Answer: B. organic chemistry

Comparative anatomy is the comparison of anatomical characteristics between different species. This includes the study of homologous and analogous structures. The comparison of DNA between species is the best way to establish evolutionary relationships. Organic chemistry has nothing to do with evolution.

TEACHER CERTIFICATION STUDY GUIDE

50. **Fossils of the dinosaur genus Saurolophus have been found in both Western North America and Mongolia. What is the most likely explanation for these findings?**
    (Rigorous)

    A. Convergent evolution on the two continents lead to two species of dinosaurs with sufficient similarity to be placed in the same genus.
    B. With the gaps in the fossil record, there are currently multiple competing theories to explain the presence of these fossils on two separate continents.
    C. A distant ancestor of the these dinosaurs evolved before these land masses became separated.
    D. Although Asia and North American were separate continents at the time, low sea levels made it possible for the dinosaurs to walk from one continent to the other.

Answer: D. Although Asia and North American were separate continents at the time, low sea levels made it possible for the dinosaurs to walk from one continent to the other.

Convergent evolution explains how different species develop similar traits but are classified in a different genus because of their greater differences. In the example provided, the fossil record indicates too many similarities, and thus the Saurolophus are a single genus of dinosaurs. (Koalas are an interesting example of convergent evolution. They are one of the few species to have distinct fingerprints like humans). Although there are gaps in the fossil record, there is one theory that is supported by significant evidence and to which most of the paleontologists ascribe. The last time the two land masses were part of the same continent (known as Laurasi) was approximately 40 million years before the existence of Saurolophus. The last and most likely possibility is that the dinosaurs walked from Asia to North America via the Bering Land Bridge (where the Bering Strait is now). The Bering Land bridge existed because large glaciers sequestered enough water to lower the level of the ocean. This lowering of ocean level was enough to expose what is now the ocean floor between Alaska and Siberia.

BIOLOGY

Skill 3.2.2 Mechanisms

Heritable variation is responsible for the individuality of organisms. An individual's phenotype is based on inherited genotype and the surrounding environment. Mutation and sexual recombination creates genetic variation. **Mutations** may be errors in replication or spontaneous rearrangements of one or more segments of DNA.

Mutations contribute a minimal amount of variation in a population. It is the unique **recombination** of existing alleles that cause the majority of genetic differences. Recombination is caused by the crossing over of the parent genes during meiosis. This results in unique offspring. With all the possible mating combinations in the world, it is obvious how sexual reproduction is the primary cause of genetic variation.

Natural selection is based on the survival of certain traits in a population through the course of time. The phrase "survival of the fittest," is often associated with natural selection. Fitness is the contribution an individual makes to the gene pool of the next generation.

Natural selection acts on phenotypes. An organism's phenotype is constantly exposed to its environment. Based on an organism's phenotype, selection indirectly adapts a population to its environment by maintaining favorable genotypes in the gene pool.

There are three modes of natural selection. **Stabilizing selection** favors the more common phenotypes, **directional selection** shifts the frequency of phenotypes in one direction, and **diversifying selection** favors individuals on both extremes of the phenotypic range.

Sexual selection leads to the secondary sex characteristics of males and females. Animals that use mating behaviors may be successful or unsuccessful. A male animal that lacks attractive plumage or has a weak mating call will not attract females, thereby eventually limiting that gene in the gene pool. Mechanical isolation, where sex organs do not fit the female, has an obvious disadvantage.

There are two theories on the rate of evolution. **Gradualism** is the theory that minor evolutionary changes occur at a regular rate. Darwin's book, "On the Origin of Species," is based on this theory of gradualism.

Charles Darwin was born in 1809 and spent 5 years in his twenties on a ship called the *Beagle*. Of all the locations Darwin visited, he became infatuated with the Galapagos Islands. There he collected 13 species of finches that were quite similar. He could not accurately determine whether these finches were of the same species. He later learned these finches were in fact separate species.

Darwin began to hypothesize that a new species arose from its ancestors by gradually collecting adaptations to a different environment. Darwin's most popular hypothesis involved the beak size of Galapagos finches. He theorized that the finches' beak sizes evolved to accommodate different food sources.

Although Darwin believed the origin of species was gradual, he was bewildered by gaps in the fossil records of living organisms. **Punctuated equilibrium** is a model of evolution stating that species form rapidly over relatively short periods of geological history, and then progress through long periods of stasis with little or no change. Punctuationalists use fossil records to support their claim. It is probable that both gradualism and punctuated equilibrium are correct, depending on the particular lineage studied.

TEACHER CERTIFICATION STUDY GUIDE

Sample Test Questions and Rationale

51. Which of the following best exemplifies the Theory of Inheritance of Acquired Traits?
    (Average Rigor)

    A. Giraffes need to reach higher for leaves to eat, so their necks stretch. The giraffe babies are then born with longer necks. Eventually, there are more long-necked giraffes in the population.
    B. Giraffes with longer necks are able to reach more leaves, so they eat more and have more babies than other giraffes. Eventually, there are more long-necked giraffes in the population.
    C. Giraffes want to reach higher for leaves to eat, so they release enzymes into their bloodstream, which in turn causes fetal development of longer-necked giraffes. Eventually, there are more long-necked giraffes in the population.
    D. Giraffes with long necks are more attractive to other giraffes, so they get the best mating partners and produce more babies. Eventually, there are more long-necked giraffes in the population.

Answer: A. Giraffes need to reach higher for leaves to eat, so their necks stretch. The giraffe babies are then born with longer necks. Eventually, there are more long-necked giraffes in the population.

The theory of inheritance of acquired traits states that the offspring of an individual will benefit from the adaptations of the parent. The stretching of the neck thus leads to longer neck offspring. Answer b best exemplifies the theory of natural selection, where an outside factor affects the chance of an individual to live and reproduce, and thus pass on their genetic material to the next generation. There is no evidence of desire creating genetic or developmental change in a fetus. Additionally there is no evidence that giraffes select mates based on neck length, however if they did this would be an example of sexual selection, an aspect of natural selection.

52. Any change that affects the sequence of nucleotides in a gene is called a(n):
    (Easy)

    A. deletion.
    B. polyploid.
    C. mutation.
    D. duplication.

Answer: C. mutation.

A mutation is an inheritable change in DNA. It may be an error in replication or a spontaneous rearrangement of one ore more segments of DNA. Deletion and duplication are types of mutations. Polyploidy is when an organism has more than two complete chromosome sets.

TEACHER CERTIFICATION STUDY GUIDE

53. The fossil record is often cited as providing evidence for the theory of punctuated equilibrium. Which of the following examples can only be explained by punctuated equilibrium and not gradualism? *(Rigorous)*

   I   Coelacanth fish (once thought extinct) have remained relatively unchanged for millions of years.
   II The sudden apearance of a large number of different soft-bodied animals around 530 million years ago.
   III 10 million year old fossils and modern ginko plants are nearly identical.
   IV Fossils of Red Deer from the Island of Jersey show a six-fold decrease in body weight over the last 6000 years.

   A. I, III
   B. II, IV
   C. I, II, III, IV
   D. None of the above

Answer: D. None of the above

Gradualism and punctuated equilibrium are not mutually exclusive. Since we are talking in terms of geological time, a rapid change can be thought to occur over a period of 1,000 years to 100,000 years or more. Items I and III appear to demonstrate a state of stasis, however it is possible that some changes cannot be observed in the fossil record. Items II and IV appear to show a period of sudden change. In the case of item II, the change may have occurred over the course of a million years or more. In the case of item IV, 6000 years can conceivably include 3000 generations (red deer mature at age 2). Therefore, in both item II and IV, the apparent sudden change may have actually occurred gradually.

Skill 3.2.3 Population genetics

Evolution currently is defined as a change in genotype over time. Gene frequencies shift and change from generation to generation. Populations evolve, not individuals. The **Hardy-Weinberg** theory of gene equilibrium is a mathematical prediction to show shifting gene patterns. Let us use the letter "A" to represent the dominant condition of normal skin pigment, and the letter "a" to represent the recessive condition of albinism. In a population, there are three possible genotypes: *AA, Aa* and *aa*. *AA* and *Aa* would have normal skin pigment and only *aa* would be albinos.

According to the Hardy-Weinberg law, there are five requirements that keep gene frequency stable and limit evolution:

1. There is no mutation in the population.
2. There are no selection pressures; one gene is not more desirable in the environment.
3. There is no mating preference; mating is random.
4. The population is isolated; there is no immigration or emigration.
5. The population is large (mathematical probability is more correct with a large sample).

The above conditions are extremely difficult to meet. If these five conditions are not met, then gene frequency can shift, leading to evolution. Let us say in a population, 75% of the population has normal skin pigment (*AA* and *Aa*) and 25% are albino (*aa*). Using the following formula, we can determine the frequency of the *A* allele and the *a* allele in a population.

This formula can be used over generations to determine if evolution is occurring. The formula is: $1 = p^2 + 2pq + q^2$; where 1 is the total population, p^2 is the number of *AA* individuals, $2pq$ is the number of *Aa* individuals, and q^2 is the number of *aa* individuals.

Since you cannot tell by looking if an individual is *AA* or *Aa*, you must use the *aa* individuals to find that frequency first. As stated above *aa* was 25% of the population. Since $aa = q^2$, we can determine the value of q (or *a*) by finding the square root of 0.25, which is 0.5. Therefore, 0.5 of the population has the *a* gene. In order to find the value for p, use the following formula: $1 = p + q$. This would make the value of $p = 0.5$.

The gene pool is all the alleles at all gene loci in all individuals of a population. The Hardy-Weinberg theorem describes the gene pool in a non-evolving population. It states that the frequencies of alleles and genotypes in a population's gene pool are random unless acted on by something other than sexual recombination.

Now, to find the number of *AA*, plug it into the first formula:

$$AA = p^2 = 0.5 \times 0.5 = 0.25$$
$$Aa = 2pq = 2(0.5 \times 0.5) = 0.5$$
$$aa = q^2 = 0.5 \times 0.5 = 0.25$$

Any question on the test dealing with the Hardy-Weinberg theorem will have an obvious squared number. The square of that number will be the frequency of the recessive gene, and you can figure anything else out knowing the formula and the frequency of q.

When frequencies vary from the Hardy-Weinberg equilibrium, the population is evolving. The change to the gene pool is on such a small scale that it is called microevolution. Certain factors increase the chances of variability in a population, thus leading to evolution. Items that increase variability include mutations, sexual reproduction, immigration, large population, and variation in geographic locale. Changes that decrease variation are natural selection, emigration, small population, and random mating.

Sample Test Questions and Rationale

54. Evolution occurs in...
    (Easy)

    A. individuals.
    B. populations.
    C. organ systems.
    D. cells.

Answer: B. populations.

Evolution is a change in genotype over time. Gene frequencies shift and change from generation to generation. Populations evolve, not individuals.

55. Which of the following factors will affect the Hardy-Weinberg law of equilibrium, leading to evolutionary change?
    (Average Rigor)

    A. no mutations
    B. non-random mating
    C. no immigration or emigration
    D. large population

Answer: B. non-random mating

There are five requirements to maintain Hardy-Weinberg equilibrium: no mutation, no selection pressures, an isolated population, a large population, and random mating.

56. **If a population is in Hardy-Weinberg equilibrium and the frequency of the recessive allele is 0.3, what percentage of the population would be expected to be heterozygous?**
    (Rigorous)

    A. 9%
    B. 49%
    C. 42%
    D. 21%

Answer: C. 42%

0.3 is the value of q. Therefore, $q^2 = 0.09$. According to the Hardy-Weinberg equation, $1 = p + q$.

$1 = p + 0.3$.
$p = 0.7$
$p^2 = 0.49$

Next, plug q^2 and p^2 into the equation $1 = p^2 + 2pq + q^2$.

$1 = 0.49 + 2pq + 0.09$ (where 2pq is the number of heterozygotes).
$1 = 0.58 + 2pq$
$2pq = 0.42$

Multiply by 100 to get the percent of heterozygotes, 42%.

57. **An animal choosing its mate because of attractive plumage or a strong mating call is an example of:**
    (Average Rigor)

    A. Sexual Selection.
    B. Natural Selection.
    C. Mechanical Isolation.
    D. Linkage

Answer: A. Sexual Selection

Sexual selection, the act of choosing a mate, allows animals to have some choice in the genetic makeup of its offspring. The answer is (A).

58. Which of the following is an example of a phenotype that gives the organism an advantage in their home environment?
(Rigorous)

    A. The color of Pepper Moths in England
    B. Thornless roses in a nature perserve
    C. Albinism in naked mole rats
    D. The large thorax of a Mediterranean fruit fly

Answer: A. The color of Pepper Moths in England

Thornless roses are not naturally occuring and would not convey an advantage against natural predators. Albinism is not any more common in naked mole rats than other species and would not be advantageous in a subterrian environment. Thorax size in Mediterranean fruit flies has been linked to sexual selection, however sexual selection is not an environmental pressure. The Pepper Moth of England is the most often cited example of natural selection. A dramatic shift in color frequency occured during the industrial revolution. This color change occurred because moths camouflage on trees and during the industrial revolution soot changed the color of trees.

Skill 3.2.4 Speciation

The most commonly used species concept is the **Biological Species Concept (BSC)**. This concept states that a species is a reproductive community of populations that occupy a specific niche in nature. It focuses on reproductive isolation of populations as the primary criterion for recognition of species status. The biological species concept does not apply to organisms that are asexual in their reproduction, fossil organisms, or distinctive populations that hybridize.

Reproductive isolation is caused by any factor that impedes two species from producing viable, fertile hybrids. Reproductive barriers can be categorized as **prezygotic** (premating) or **postzygotic** (postmating).

The prezygotic barriers include:

1. Habitat isolation – species occupy different habitats in the same territory.
2. Temporal isolation – populations reaching sexual maturity/flowering at different times of the year.
3. Ethological isolation – behavioral differences that reduce or prevent interbreeding between individuals of different species (including pheromones and other attractants).
4. Mechanical isolation – structural differences that make gamete transfer difficult or impossible.

5. Gametic isolation – male and female gametes do not attract each other; no fertilization.

The postzygotic barriers include:

1. Hybrid inviability – hybrids die before sexual maturity.
2. Hybrid sterility – disrupts gamete formation; no normal sex cells.
3. Hybrid breakdown – reduces viability or fertility in progeny of the F_2 backcross.

Geographical isolation can also lead to the origin of species. **Allopatric speciation** is speciation without geographic overlap. It is the accumulation of genetic differences through division of a species' range, either through a physical barrier separating the population or through expansion by dispersal. In **sympatric speciation**, new species arise within the range of parent populations. Populations are sympatric if their geographical range overlaps. This usually involves the rapid accumulation of genetic differences (usually chromosomal rearrangements) that prevent interbreeding with adjacent populations.

Sample Test Questions and Rationale

59. **Members of the same species:**
    (Easy)

    A. look identical.
    B. never change.
    C. reproduce successfully within their group.
    D. live in the same geographic location.

Answer: C. reproduce successfully within their group.

Species are defined by the ability to successfully reproduce with members of their own kind.

60. Reproductive isolation results in...
    (Average Rigor)

    A. extinction.
    B. migration.
    C. follilization.
    D. speciation.

Answer: D. speciation.

Reproductive isolation is caused by any factor that impedes two species from producing viable, fertile hybrids. Reproductive isolation of populations is the primary criterion for recognition of species status.

61. The biological species concept applies to
    (Average Rigor)

    A. asexual organisms.
    B. extinct organisms.
    C. sexual organisms.
    D. fossil organisms.

Answer: C. sexual organisms.

The biological species concept states that a species is a reproductive community of populations that occupy a specific niche in nature. It focuses on reproductive isolation of populations as the primary criterion for recognition of species status. The biological species concept does not apply to organisms that are completely asexual in their reproduction, fossil organisms, or distinctive populations that hybridize.

Skill 3.2.5   Phylogeny

The typical graphic representation of classification is a **phylogenetic tree**, which represents a hypothesis of the relationships based on branching of lineages through time within a group.

Every time you see a phylogenetic tree, you should be aware that it is making statements on the degree of similarity between organisms, or the particular pattern in which the various lineages diverged (phylogenetic history).

Cladistics is the study of phylogenetic relationships of organisms by analysis of shared, derived character states. Cladograms are constructed to show evolutionary pathways. Character states are polarized in cladistic analysis to be plesiomorphous (ancestral features), symplesiomorphous (shared ancestral features), apomorphous (derived features), and synapomorphous (shared, derived features).

Skill 3.2.6   Origin of life

The hypothesis that life developed on Earth from nonliving materials is the most widely accepted theory. The transformation from nonliving materials to life had four stages. The first stage was the nonliving (abiotic) synthesis of small monomers such as amino acids and nucleotides. In the second stage, these monomers combine to form polymers, such as proteins and nucleic acids. The third stage was the formation of protobionts, droplets containing proteins or nucleic acids surrounded by a membrane-like structure. The last stage was the origin of heredity, with RNA as the first genetic material.

The first stage of this theory was hypothesized in the 1920s. A. I. Oparin and J. B. S. Haldane were the first to theorize that the primitive atmosphere was a reducing atmosphere without oxygen. The gases were rich in hydrogen, methane, water, and ammonia. In the 1950s, Stanley Miller proved Oparin's theory in the laboratory by combining the above gases. When given an electrical spark, he was able to synthesize simple amino acids. It is commonly accepted that amino acids appeared before DNA. Other laboratory experiments have also supported the other stages in origin of life theory.

Other scientists believe simpler hereditary systems originated before nucleic acids. In 1991, Julius Rebek was able to synthesize a simple organic molecule that could replicate itself. According to his theory, this simple molecule represents a precursor of RNA.

Prokaryotes are the simplest life form. Their small genome size limits the number of genes that control metabolic activities. Over time, some prokaryotic groups became multicellular organisms for this reason. Prokaryotes then evolved to form complex bacterial communities in which each species benefited from one another.

The **endosymbiotic theory** of the origin of eukaryotes states that eukaryotes arose from symbiotic groups of prokaryotic cells. According to this theory, smaller prokaryotes lived within larger prokaryotic cells, eventually evolving into chloroplasts and mitochondria.

Chloroplasts are the descendant of photosynthetic prokaryotes and mitochondria are likely the descendants of bacteria that were aerobic heterotrophs. Serial endosymbiosis is a sequence of endosymbiotic events. Serial endosymbiosis may also play a role in the progression of life forms to become eukaryotes.

Sample Test Questions and Rationale

62. The first cells that evolved on earth were probably of which type?
    (Easy)

    A. autotrophs
    B. eukaryotes
    C. heterotrophs
    D. prokaryotes

Answer: D. prokaryotes

Prokaryotes were first observed in the fossil record 3.5 billion years ago. Their ability to adapt to the environment allows them to thrive in a wide variety of habitats.

63. All of the following gasses made up the primitive atmosphere except...
    (Average Rigor)

    A. ammonia
    B. methane
    C. oxygen
    D. hydrogen

Answer: C. Oxygen

In the 1920's, Oparin and Haldane were the first to theorize that the primitive atmosphere was a reducing atmosphere with no oxygen. The atmosphere was rich in hydrogen, methane, water, and ammonia.

Skill 3.2.7 Species extinction

Extinction the ceasing of existence of a species resulting in reduced biodiversity. The point of extinction is generally considered as the death of the last species of taxa. However, determining the exact moment of extinction is not easy.

In the context of evolution, new species are created by speciation – where new varieties of organisms arise and thrive when they are able to find and exploit an ecological niche. Species become extinct when they are no longer able to survive in changing conditions or against superior domination. A typical species becomes extinct within 10 million years of its first appearance, although some species, called "living fossils" survive virtually unchanged for hundreds of millions of years

Prior to the existence of human beings across the earth, extinction was a purely natural phenomenon that occurred at a low rate. Mass extinction was very rare. Within the last 100,000 years, as the numbers of human beings have continued to grow and expand, species extinction has increased at an unprecedented rate.

A species becomes extinct when the last existing member of that species dies. Extinction therefore becomes a certainty when there are no surviving individuals that are able to reproduce and create a new generation. A species may become functionally extinct when only a handful of individuals survive, which are unable to reproduce due to poor health, age, sparse distribution, or other reasons.

It is important to note that humans have also made significant attempts to preserve critically endangered species through the creation of the conservation status.

TEACHER CERTIFICATION STUDY GUIDE

DOMAIN 4.0 DIVERSITY OF LIFE, PLANTS, AND ANIMALS

Competency 4.1 Diversity of life

Skill 4.1.1 Classification schemes and the five-kingdom system

Scientists estimate that there are more than ten million different species of living things. Of these, 1.5 million have been named and classified. Systems of classification show similarities and provide scientists with a worldwide system of organization.

Carolus Linnaeus is known as the father of taxonomy. **Taxonomy** is the science of classification. Linnaeus based his system on morphology (study of structure). Later on, evolutionary relationships (phylogeny) were also used to sort and group species. The modern classification system uses binomial nomenclature, a two-word name for every species. The genus is the first part of the name and the species is the second part. Notice in the example below that *Homo sapiens* is the scientific name for humans. Starting with the kingdom, the groups get smaller and more alike as one moves down to the level of species.

Kingdom: Animalia, Phylum: Chordata, Subphylum: Vertebrata, Class: Mammalia, Order: Primate, Family: Hominidae, Genus: Homo, Species: sapiens

Species are defined by their ability to successfully reproduce with other members of their species.

Several different morphological criteria are used to classify organisms:

1. **Ancestral characters** - characteristics that are unchanged after evolution (e.g. 5 digits on the hand of an ape).
2. **Derived characters** - characteristics that have evolved more recently (e.g. the absence of a tail on an ape).
3. **Conservative characters** - traits that change slowly.
4. **Homologous characters** - characteristics with the same genetic basis but used for a different function. (e.g., wing of a bat, arm of a human. The bone structure is the same, but the limbs are used for different purposes).
5. **Analogous characters** – structures that differ but are used for similar purposes (e.g. the wing of a bird and the wing of a butterfly).
6. **Convergent evolution** - development of similar adaptations by organisms that are unrelated.

Biological characteristics are also used to classify organisms. Protein comparison, DNA comparison, and analysis of fossilized DNA are powerful comparative methods used to measure evolutionary relationships between species. Taxonomists consider the organism's life history, biochemical (DNA) makeup, behavior, and geographical distribution. The fossil record is also used to show evolutionary relationships.

The current five-kingdom system separates prokaryotes from eukaryotes. The prokaryotes belong to the Kingdom Monera while the eukaryotes belong to Kingdoms Protista, Plantae, Fungi, or Animalia. Recent comparisons of nucleic acids and proteins between different groups of organisms have revealed discrepancies in the five-kingdom system. Based on these comparisons, alternative kingdom systems have emerged. Six and eight kingdom systems as well as a three-domain system have been proposed as more accurate classification systems. It is important to note that classification systems evolve as more information regarding characteristics and evolutionary histories of organisms arise.

Sample Test Questions and Rationale

64. Man's scientific name is Homo sapiens. Choose the proper classification beginning with kingdom and ending with order. *(Average Rigor)*

    A. Animalia, Vertebrata, Mammalia, Primate, Hominidae
    B. Animalia, Vertebrata, Chordata, Homo, sapiens
    C. Animalia, Chordata, Vertebrata, Mammalia, Primate
    D. Chordata, Vertebrata, Primate, Homo, sapiens

Answer: C. Animalia, Chordata, Vertebrata, Mammalia, Primate

The order of classification for humans is as follows: Kingdom, Animalia; Phylum, Chordata; Subphylum, Vertebrata; Class, Mammalia; Order, Primate; Family, Hominadae; Genus, Homo; Species, sapiens.

TEACHER CERTIFICATION STUDY GUIDE

65. The two major ways to determine taxonomic classification are: *(Average Rigor)*

    A. evolution and phylogeny
    B. reproductive success and evolution
    C. phylogeny and morphology.
    D. size and color.

Answer: C. phylogeny and morphology.

Taxonomy is based on structure (morphology) and evolutionary relationships (phylogeny).

66. The scientific name *Canis familiaris* refers to the animal's: *(Easy)*

    A. kingdom and phylum names.
    B. genus and species names.
    C. class and species names.
    D. order and family names.

Answer: B. genus and species names.

Each species is scientifically known by a two-part name, a system called binomial nomenclature. The first word in the name is the genus and the second word is its specific epithet (species name).

Skill 4.1.2   Characteristics and representatives of kingdoms

The traditional classification of living things is the five-kingdom system. The five kingdoms are Monera, Protista, Fungi, Plantae, and Animalia. The following is a comparison of the cellular characteristics of members comprising the five kingdoms.

Kingdom Monera

Members of the Kingdom Monera are single-celled, prokaryotic organisms. Like all prokaryotes, Monerans lack nuclei and other membrane bound organelles, but do contain circular chromosomes and ribosomes. Most Monerans possess a cell wall made of peptidoglycan, a combination of sugars and proteins. Some Monerans also possess capsules and external motility devices (e.g. pili or flagella). The Kingdom Monera includes both eubacteria and archaebacteria. Though archaebacteria are structurally similar to eubacteria in many ways, there are key differences such as cell wall composition (archae lack peptidoglycan).

Kingdom Protista

Protists are eukaryotic, usually single-celled organisms (though some protists are multicellular). The Kingdom Protista is very diverse, containing members with characteristics of plants, animals, and fungi. All protists possess nuclei and some types of protists possess multiple nuclei. Most protists contain many mitochondria for energy production, and photosynthetic protists contain specialized structures called plastids where photosynthesis occurs. Motile protists possess external cilia or flagella. Finally, many protists have cell walls that do not contain cellulose.

Kingdom Fungi

Fungi are eukaryotic organisms that are mostly multicellular (single-celled yeast are the exception). Fungi possess cell walls composed of chitin. Fungal organelles are similar to animal organelles. Fungi are non-photosynthetic and possess neither chloroplasts nor plastids. Many fungal cells, like animal cells, possess centrioles. Fungi are also non-motile and release exoenzymes into the environment to dissolve food.

Kingdom Plantae

Plants are eukaryotic, multicellular, and have square-shaped cells. Plant cells possess rigid cell walls composed mostly of cellulose. Plant cells also contain chloroplasts and plastids for photosynthesis. Plant cells generally do not possess centrioles. Another distinguishing characteristic of plant cells is the presence of a large, central vacuole that occupies 50-90% of the cell interior. The vacuole stores acids, sugars, and wastes. Because of the presence of the vacuole, the cytoplasm is limited to a very small part of the cell.

Kingdom Animalia

Animals are eukaryotic, multicellular, and motile. Animal cells do not possess cell walls or plastids, but do possess a complex system of organelles. Most animal cells also possess centrioles, microtubule structures that play an important role in spindle formation during replication.

The three-domain system of classification, introduced by Carl Woese in 1990, emphasizes the separation of the two types of prokaryotes. The following is a comparison of the cellular characteristics of members of the three domains of living organisms: Eukarya, Bacteria, and Archaea.

Domain Eukarya

The Eukarya domain includes all members of the protist, fungi, plant, and animal kingdoms. Eukaryotic cells possess a membrane bound nucleus and other membranous organelles (e.g., mitochondria, Golgi, ribosomes). The chromosomes of Eukarya are linear and usually complex with histones (protein spools). The cell membranes of eukaryotes consist of glycerol-ester lipids and sterols. The ribosomes of eukaryotes are 80 Svedburg (S) units in size. Finally, the cell walls of those eukaryotes that have them (i.e., plants, algae, fungi) are polysaccharide in nature.

Domain Bacteria

Prokaryotic members of the Kingdom Monera, not classified as Archaea, are members of the Bacteria domain. Bacteria lack a defined nucleus and other membranous organelles. The ribosomes of bacteria measure 70 S in size. The chromosome of Bacteria is usually a single, circular molecule that does not complex with histones. The cell membranes of Bacteria lack sterols and consist of glycerol-ester lipids. Finally, most Bacteria possess a cell wall made of peptidoglycan.

Domain Archaea

There are three kinds of organisms with archaea cells: **methanogens**, obligate anaerobes that produce methane, **halobacteria**, which can live only in concentrated brine solutions, and **thermoacidophiles**, which can live only in acidic hot springs. Members of the Archaea domain are prokaryotic and similar to bacteria in most aspects of cell structure and metabolism. However, transcription and translation in Archaea are similar to the processes of eukaryotes, not bacteria. In addition, the cell membranes of Archaea consist of glycerol-ether lipids in contrast to the glycerol-ester lipids of eukaryotic and bacterial membranes. Finally, the cell walls of Archaea are not made of peptidoglycan, but consist of other polysaccharides, protein, and glycoprotein.

TEACHER CERTIFICATION STUDY GUIDE

Sample Test Questions and Rationale

67. **Thermoacidophiles are...**
    (Average Rigor)

    A. prokaryotes.
    B. eukaryotes.
    C. protists.
    D. archaea.

Answer: D. archaea.

Thermoacidophiles, methanogens, and halobacteria are members of the archaea group.

68. **Protists are classified into major groups according to**
    (Average Rigor)

    A. their method of obtaining nutrition.
    B. reproduction.
    C. metabolism.
    D. their form and function.

Answer: D. their form and function.

The extreme variation in protist classification reflects their diverse form, function, and life style. The protists are often grouped as algae (plant-like), protozoa (animal-like), or fungus-like based on their similarity of characteristics.

69. **All of the following are examples of a member of Kingdom Fungi except:**
    (Easy)

    A. mold.
    B. algae.
    C. mildew.
    D. mushrooms.

Answer: B. algae.

Mold, mildew, and mushrooms are all fungi. Brown and golden algae are members of the Kingdom Protista and green algae are members of the Plant Kingdom.

BIOLOGY 93

70. Which kingdom is comprised of organisms made of one cell with no nuclear membrane?
   (Easy)

   A. Monera
   B. Protista
   C. Fungi
   D. Algae

Answer: A. Monera

Monera is the only kingdom comprising unicellular organisms lacking a nucleus. Algae are classified as a protist. Algae may be uni- or multicellular and have a nucleus.

71. Within the Phylum Mollusca, there are examples of both open and closed circulatory systems. Which of the following is a feature that is not common to both the open and closed cirulatory systems of molluscs?
   (Rigorous)

   A. Hemocoel
   B. Plasma
   C. Vessels
   D. Heart

Answer: A. Hemocoel

Hemocoel is the blood filled cavity that is present in animals with an open circulatory system. Unlike some other open circulatory systems, the molluscs have three blood vessels, two to bring blood from the lungs and one to push blood into the hemocoel.

72. Which of the following systems considers Archea (or Archeabacteria) as the most inclusive level of the taxonomic system

   I   Three Domain System
   II Five Kingdom System
   III Six Kingdom System
   IV Eight Kingdom System
   (Rigorous)

   A. II, III
   B. I, IV
   C. I, III, IV
   D. I, II, III, IV

Answer: C. I, III, IV

In the five kingdom system the subkingdom Archaebacteriobionta is under the kingdom Monera.

73. Laboratory researchers distinguish and classify fungi from plants because the cell walls of fungi contain _____.
    (Rigorous)

   A. chitin.
   B. lignin.
   C. lipopolysaccharides.
   D. cellulose.

Answer: A. contain chitin.

All of the possible answers are compounds found in cell walls. Cellulose is found in the cell wall of all plants, while lignin is found only in the cell wall of vascular plants. Lipopolysaccharides are found in the cell wall of gram negative bacteria. Chitin is the only compound uniquely found in fungal cell walls.

Competency 4.2 Plants

Skill 4.2.1   Evolution (including adaptation to land and major divisions)

Plants require adaptations that allow them to absorb light for photosynthesis. Since they are unable to move about, they must evolve methods to allow them to reproduce successfully. As time passed, the plants moved from a water environment to the land. Advantages of life on land included more available light and a higher concentration of carbon dioxide. Originally, there were no predators and less competition for space on land. Plants had to evolve methods of support, reproduction, respiration, and conservation of water once they moved to land. Plant reproduction occurs through an alternation of generations meaning a haploid stage in the plant's life history alternates with a diploid stage. Specific plant tissues evolved in order to obtain water and minerals from the earth. The plant's wax cuticle prevents the loss of water while the leaves capture light and carbon dioxide for photosynthesis. Stomata provide openings on the underside of leaves for oxygen to move in or out of the plant and for carbon dioxide to move in. Roots evolved to provide a method of anchorage and the polymer lignin evolved to provide structural support.

The **non-vascular plants** represent an evolutionary grade characterized by several primitive features: lack of roots, lack of conducting tissues, reliance on absorption of water that falls on the plant or condenses on the plant in high humidity, and a lack of leaves. Non-vascular plants include the liverworts, hornworts, and mosses. Each is recognized as a separate division.

The characteristics of **vascular plants** are as follows: they contain lignin which provides rigidity and strength to cell walls for upright growth, tracheid cells for water transport and sieve cells for nutrient transport, and underground stems (rhizomes) as a structure from which adventitious roots originate.

There are two kinds of vascular plants: non-seeded and seeded. The non-seeded vascular plant divisions include Division Lycophyta (club mosses), Division Sphenophyta (horsetails), and Division Pterophyta (ferns). The seeded vascular plants differ from the non-seeded plants by their method of reproduction, which we will discuss later. Vascular seed plants are divided into two groups, the gymnosperms and the angiosperms.

Gymnosperms were the first plants to evolve with the use of seeds for reproduction, which made them less dependent on water to assist in reproduction. Their seeds and the pollen from the male are carried by the wind. Gymnosperms have cones that protect the seeds. Gymnosperm divisions include Division Cycadophyta (cycads), Division Ginkgophyta (ginkgo), Division Gnetophyta (gnetophytes), and Division Coniferophyta (conifers).

Angiosperms are the largest group in the plant kingdom. They are the flowering plants and produce true seeds for reproduction. Angiosperms arose about seventy million years ago when dinosaurs were disappearing. The land was drying up and the plants' ability to produce seeds that could remain dormant until conditions became acceptable allowed for their success. They also have more advanced vascular tissue and larger leaves for increased photosynthesis. Angiosperms consist of only one division, the Anthrophyta. Angiosperms are divided into monocots and dicots. Monocots have one cotelydon (seed leaf) and parallel veins on their leaves. Their flower petals are in multiples of threes. Dicots have two cotelydons and branching veins on their leaves. Flower petals are in multiples of fours or fives.

Sample Test Questions and Rationale

74.

Identify the correct characteristics of the plant pictured above. *(Rigorous)*

    A. seeded, non-vascular
    B. non-seeded, vascular
    C. non-seeded, non-vascular
    D. seeded, vascular

Answer: B. non-seeded, vascular

The picture above is of a fern, Division Pterophyta, which is a spore bearing vascular plant.

75. Which of the following is a characteristic of a monocot?
    (Rigorous)

    A. parallel veins in leaves
    B. flower petals occur in multiples of 4 or 5
    C. two seed leafs
    D. vascular tissue absent from the stem

Answer: A. parallel veins in leaves

Monocots have one cotelydon, parallel veins in their leaves, and their flower petals are in multiples of threes. Dicots have flower petals in multiples of fours and fives.

76. Spores are the reproductive mode for which of the following group of plants?
    (Average Rigor)

    A. algae
    B. flowering plants
    C. conifers
    D. ferns

Answer: D. ferns

Ferns are non-seeded vascular plants. All plants in this group have spores and require water for reproduction. Algae, flowering plants, and conifers are not in this group of plants.

Skill 4.2.2 Anatomy (including roots, stems, leaves, and reproductive structures)

Roots, stems, leaves, and reproductive structures are the most functionally important parts of plant anatomy. Different types of plants have distinctive anatomical structures. Thus, a discussion of plant anatomy requires an understanding of the classifications of plants.

Roots absorb water and minerals and exchange gases in the soil. Like stems, roots contain xylem and phloem. Xylem transports water and minerals, called xylem sap, upward. The sugar produced by photosynthesis goes down the phloem in phloem sap, traveling to the roots and other non-photosynthetic parts of the plant. In addition to water and mineral absorption, roots anchor plants in place preventing erosion by environmental conditions.

Stems are the major support structure of plants. Stems consist primarily of three types of tissue; dermal, ground, and vascular. Dermal tissue covers the outside surface of the stem to prevent excessive water loss and control gas exchange. Ground tissue consists mainly of parenchyma cells and surrounds the vascular tissue providing support and protection. Finally, vascular tissue, xylem and phloem, provides long distance transport of nutrients and water. Leaves enable plants to capture light and carbon dioxide for photosynthesis. Photosynthesis occurs primarily in the leaves. Plants exchange gases through their leaves via stomata, small openings on the underside of the leaves. Stomata allow oxygen to move in or out of the plant and carbon dioxide to move in. Leaf size and shape varies greatly between species of plants and botanists often identify plants by their characteristic leaf patterns.

Reproductive Structures

The sporophyte (diploid) generation is the dominant phase in plant reproduction and makes up almost their whole life cycle. Sporophytes contain a diploid set of chromosomes and form haploid spores by meiosis. Spores develop into gametophytes (the haploid generation) that produce male or female gametes by mitosis. Finally, the male and female gametes fuse producing a diploid zygote that develops into a new sporophyte.

Angiosperm reproductive structures are the flowers.

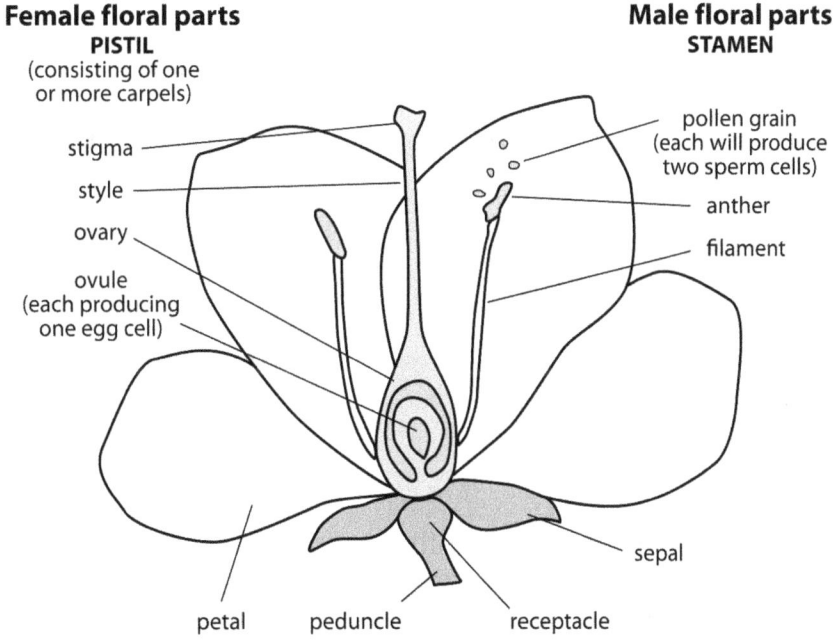

The male gametophytes are pollen grains and the female gametophytes are embryo sacs that are inside of the ovules. The male pollen grains form in the anthers at the tips of the stamens. The ovaries contain the female ovules. Finally, the stamen is the reproductive organ of the male and the carpel is the reproductive organ of the female.

Sample Test Question and Rationale

77.

Using the following taxonomic key identify the tree that this branch came from?

1 - Are the leaves PALMATELY COMPOUND (BLADES arranged like fingers on a hand)? – go to question 2
1 - Are the leaves PINNATELY COMPOUND (BLADES arranged like the vanes of a feather)? – go to question 3

2 - Are there usually 7 BLADES - Aesculus hippocastanum
2 - Are there usually 5 BLADES - Aesculus glabra

3 - Are there mostly 3-5 BLADES that are LOBED or coarsely toothed? - Acer negundo
3 - Are there mostly 5-13 BLADES with smooth or toothed edges? - Fraxinus Americana
(Rigorous)

    A. Aesculus hippocastanum
    B. Aesculus glabra
    C. Acer negundo
    D. Fraxinus Americana

Answer: C. Acer negundo

The leaves are pinnately compound, with 5 coarsly toothed leaves, leading to the answer: Acer negundo. The list below includes the scientific name and the common name for the all the plants listed above:
Aesculus hippocastanum (Horsechestnut)
Aesculus glabra (Ohio Buckeye)
Acer negundo (Boxelder, Ashleaf Maple)
Fraxinus Americana (White Ash)

Skill 4.2.3 Physiology (C3 and C4 photosynthesis, hormones, photoperiods, water and nutrient uptake, translocation)

For a general overview of photosynthesis, see **Skill 2.1.7**.

C3 and C4 Photosynthesis

While photosynthesis is common to all plant species, the specific mechanism often differs. Two major types of photosynthesis are C3 and C4. C3 photosynthesis is the typical mechanism of photosynthesis that most plants use. C4 photosynthesis, on the other hand, utilizes adaptations to enable a more efficient use of water in arid environments. C3 and C4 photosynthesis differ in several key ways.

In C3 photosynthesis, the first carbon compound formed from carbon dioxide contains three carbon atoms. C3 plants use a single enzyme, ribulosodiphosphatcarboxylase (RUBISCO), to collect carbon dioxide from the air and carry out photosynthesis. Finally, C3 photosynthesis takes place in cells throughout the leaf.

In C4 photosynthesis, the first carbon compound formed from carbon dioxide contains four carbon atoms. In contrast to C3 plants, C4 plants use two enzymes for photosynthesis. One enzyme, phosphoenolpyruvate carboxylase (PEP carboxylase), collects carbon dioxide and delivers it directly to the second enzyme, RUBISCO. This adaptation increases the rate of photosynthesis in high light intensity and high temperature conditions. In addition, unlike in C3 plants, C4 photosynthesis only occurs in internal leaf cells. Thus, C4 adaptations increase the efficiency of photosynthesis in dry conditions and minimize water loss. Faster and more efficient uptake and delivery of carbon dioxide allows the plant to keep its stomata closed longer, limiting water loss, while still acquiring adequate amounts of carbon dioxide.

Hormones

Plant hormones, or plant growth regulators, are chemicals secreted internally that regulate growth and development. Plant hormones are present in low concentrations, produced in specific locations, and often act on cells at other locations. The mechanism of hormonal signaling involves attachment of hormone molecules to protein receptors, transmission of the signal along a transduction pathway, and the activation of particular genes. The five major classes of plant growth regulators are: (1) auxins, (2) abscisic acid, (3) gibberellins, (4) ethylene, and (5) cytokinins.

Auxins play a major role in many growth and behavioral processes. Auxins promote cell elongation in growing shoot tips and cell expansion in swelling roots and fruits. In addition, auxins promote apical dominance, the tendency of the main stem of plants to grow more strongly than the side stems. Other effects of auxins include phototropism (plants bending toward light), stimulation of ethylene synthesis, stimulation of cell division, and inhibition of abscission (i.e. leaf shedding).

Abscisic acid (ABA) plays an important role in promoting dormancy, inhibiting growth, and responding to water stress. High levels of ABA induce seed dormancy and inhibit germination. In addition, ABA build up during water stress promotes closing of stomata (i.e. pores). Finally, ABA prevents fruit ripening.

Gibberellins have a wide range of effects. The most important effect of gibberellins is stem elongation. Gibberellins also promote flower and fruit formation. Finally, gibberellins stimulate the growth and development of seeds.

Ethylene is the major hormone involved in fruit ripening and abscission. Ethylene also induces seed germination, root hair growth, and flowering.

Cytokinins play a key role in cell division. Cytokinins generally promote shoot development and inhibit root development. In addition, cytokinins delay senescence in flowers and fruits, promote chlorophyll production, and promote photosynthesis.

Photoperiods

A photoperiod is the duration of a plant's daily exposure to light. Variations in the length of photoperiods effect the growth, development, and physiological processes of plants. For example, plants adapt to seasonal changes in photoperiods by increasing or decreasing growth processes and changing patterns of photosynthesis and respiration. In addition, over time, species of plants develop traits that allow them to thrive in the characteristic photoperiods of their native environment.

Water and nutrient uptake and translocation

Roots absorb water and minerals and exchange gases in the soil. The xylem transports water and minerals, called xylem sap, upwards. This is pulled upwards in a process called **transpiration**. Transpiration is the evaporation of water from leaves. Gases are exchanged through the leaves and photosynthesis occurs. The sugar produced by photosynthesis goes down the phloem in phloem sap. This sap is transported to the roots and other non-photosynthetic parts of the plant.

TEACHER CERTIFICATION STUDY GUIDE

Sample Test Questions and Rationale

78. Which of the following is not a factor that affects the rate of both photosynthesis and respiration in plants?
    (Average Rigor)

    A. the concentration of NADP and FAD
    B. the temperature
    C. the structure of plants
    D. the availability of different substrates

Answer: C. the structure of plants

The structure of the plants leaf affects its ability to absorb light which affects the rate of photosynthesis but not the rate of respiration.

79. Oxygen is given off in the:
    (Easy)

    A. light reaction of photosynthesis.
    B. dark reaction of photosynthesis.
    C. Krebs cycle.
    D. reduction of NAD^+ to NADH.

Answer: A. light reaction of photosynthesis.

The conversion of solar energy to chemical energy occurs in light reactions. As chlorophyll absorbs light, electrons are transferred and cause water to split, releasing oxygen as a waste product.

80. The most ATP is generated through...
    (Rigorous)

    A. fermentation.
    B. glycolysis.
    C. chemiosmosis.
    D. the Krebs cycle.

Answer: C. chemiosmosis.

The electron transport chain uses electrons to pump hydrogen ions across the mitochondrial membrane. This ion gradient is used to form ATP in a process called chemiosmosis. ATP is generated by the removal of hydrogen ions from NADH and $FADH_2$. This yields 34 ATP molecules.

BIOLOGY

81. **Which of the following is not employed by a young cactus to survive in an arid environment?**
    (Rigorous)

    A. Stem as the principle site of photosynthesis.
    B. A deep root system to reach additional sources of groundwater.
    C. CAM cycle photosynthesis.
    D. Spherical growth form.

Answer: B. A deep root system to reach additional sources of groundwater.

A waxy sperical stem as the site of photosynthesis is an adaptation that limits water loss and allows for maximum water storage. CAM cycle photosynthesis allows for the plant to open its stomata at night thus limiting possible water loss due to evaporation. Some cacti will develop a taproot when it is necessary to stabilize the plant.

82. **Oxygen created in photosynthesis comes from the breakdown of**
    (Average Rigor)

    A. carbon dioxide.
    B. water.
    C. glucose.
    D. carbon monoxide.

Answer: B. water.

In photosynthesis, water is split; the hydrogen atoms are pulled to carbon dioxide which is taken in by the plant and ultimately reduced to make glucose. The oxygen from water is given off as a waste product.

83. **A plant cell is placed in salt water. The resulting movement of water out of the cell is called...**
    (Average Rigor)

    A. facilitated diffusion.
    B. diffusion.
    C. transpiration.
    D. osmosis.

Answer: B. diffusion.

Osmosis is simply the diffusion of water across a semi-permeable membrane. Water will diffuse out of the cell if less water is present on the outside than inside the cell.

Skill 4.2.4 Reproduction (alternation of generations, fertilization and zygote formation, dispersal, germination, growth and differentiation, vegetative propagation)

Alternation of generations, fertilization, zygote formation, and dispersal

Reproduction by plants is accomplished through alternation of generations. Simply stated, a haploid stage in the plants life history alternates with a diploid stage. The diploid sporophyte divides by meiosis to reduce the chromosome number to the haploid gametophyte generation. The haploid gametophytes undergo mitosis to produce gametes (sperm and eggs). Finally, the haploid gametes fertilize to return to the diploid sporophyte stage.

Non-vascular plants need water to reproduce. The vascular, non-seeded plants reproduce with spores and also need water to reproduce. Gymnosperms use seeds for reproduction and do not require water.

Angiosperms are the most numerous and are therefore the main focus of reproduction in this section. In a process called **pollination**, plant anthers release pollen grains, which are carried by animals or the wind to plant carpels. The sperm is released to fertilize the eggs. Angiosperms reproduce through a method of double fertilization. Two sperm fertilize an ovum. One sperm produces the new plant and the other forms the food supply for the developing plant (endosperm). The ovule develops into a seed and the ovary develops into a fruit. Then the wind or animals carry the seeds to new locations and new plants form in a process called **dispersal**.

The development of the egg to form a plant occurs in three stages: growth, morphogenesis (the development of form) and cellular differentiation (the acquisition of a cell's specific structure and function).

Germination

Germination describes the initiation of growth of an organism from a resting stage. In plants, a seedling sprouts or germinates from a seed. Plant seeds contain a stored embryo and food reserves. Upon germination, the plant embryo resumes growth and continues to develop into a mature plant. Most types of plant seeds germinate only when environmental conditions, such as light, temperature, and moisture are optimal. In addition, cellular plant hormones direct the germination and growth processes.

Vegetative propagation

Vegetative propagation refers to the ability of some types of plants to reproduce asexually by producing new plants from existing vegetative structures. Examples of vegetative propagation include the formation of new plants from long underground stems, root sprouting, and budding from leaf edges. Natural vegetative propagation is most common in perennial plants. Vegetative propagation is a form of cloning, because the offspring plants are identical to the parent. In addition, humans often take advantage of the regenerative nature of plants to simplify plant breeding and propagation. Examples of man-made vegetative propagation are cuttings and graftings.

Sample Test Questions and Rationale

84. **Double fertilization refers to which of the following?**
    (Average Rigor)

    A. two sperm fertilizing one egg
    B. fertilization of a plant by gametes from two separate plants
    C. two sperm enter the plant embryo sac; one sperm fertilizes the egg, the other forms the endosperm
    D. the production of non-identical twins through fertilization of two separate eggs

Answer: C. two sperm enter the plant embryo sac; one sperm fertilizes the egg, the other forms the endosperm

In angiosperms, double fertilization is when an ovum is fertilized by two sperm. One sperm produces the new plant and the other forms the food supply for the developing plant (endosperm).

85. **The process in which pollen grains are released from the anthers is called:**
    (Easy)

    A. pollination.
    B. fertilization.
    C. blooming.
    D. dispersal.

Answer: A. pollination.

Pollen grains are released from the anthers during pollination and are carried by the wind and animals to the carpels.

86. **Which of the following is the correct order of the stages of plant development from egg to adult plant?**
    (Average Rigor)

    A. morphogenesis, growth, and cellular differentiation
    B. cell differentiation, growth, and morphogenesis
    C. growth, morphogenesis, and cellular differentiation
    D. growth, cellular differentiation, and morphogenesis

Answer: C. growth, morphogenesis, and cellular differentiation

The development of the egg to form a plant occurs in three stages: growth; morphogenesis; and cellular differentiation, the acquisition of a cell's specific structure and function.

87. **In angiosperms, the food for the developing plant is found in which of the following structures?**
    (Average Rigor)

    A. ovule
    B. endosperm
    C. male gametophyte
    D. cotyledon

Answer: B. Endosperm

The endosperm is a product of double fertilization. It is the food supply for the developing plant.

88. **In a plant cell, telophase is described as…**
    (Rigorous)

    A. the time of chromosome doubling.
    B. cell plate formation.
    C. the time when crossing over occurs.
    D. cleavage furrow formation.

Answer: B. cell plate formation.

In a plant cell, a cell plate forms during telophase. In an animal cell, a cleavage furrow forms during telophase.

89. **Which of the following is a disadvantage of budding compared to sexual reproduction?**
    (Rigorous)

    A. limited number of offspring
    B. inefficient
    C. limited genetic diversity
    D. expensive to the parent organism

Answer: C. limited genetic diversity

Budding results in an identical copy of the parent, therefore the genetic diversity of the individuals in an area is very limited. It is possible for an organism that reproduces by budding to produce a much larger number of offspring than an organism reproducing by sexual reproduction.

Competency 4.3   Animals

Skill 4.3.1   Evolution (phylogeny and classification, major phyla)

Porifera - sponges; they contain spicules for support. Porifera possess flagella for movement in the larval stage, but later become sessile and attach to a firm object. Sponges may reproduce sexually (either by cross or self-fertilization) or asexually (by budding). Porifera are filter feeders and digest food by phagocytosis. They require water to support their hydroskeleton and are therefore mostly aquatic.

Cnidaria (Coelenterata) - jellyfish; these animals possess stinging cells called a nematocyst. They may be found in a sessile polyp form with the tentacles at the top or in a moving medusa form with the tentacles floating below. Jellyfish have a hydroskeleton that requires water for support. They have no true muscles. Cnidaria may reproduce asexually (by budding) or sexually. They are the first to possess a primitive nervous system.

Platyhelminthes - flatworms; the flat shape of these animals aid in the diffusion of gases. They are the first group with true muscles. Flatworms can reproduce asexually (by regeneration) or sexually. Platyhelminthes may be hermaphroditic, possessing both sex organs but cannot fertilize themselves. These worms are parasites because they have no true nervous system.

Nematoda - roundworms; the first animal with a true digestive system with a separate mouth and anus. Roundworms may be parasites or simple consumers. Nematoda reproduce sexually with male and female worms. They possess longitudinal muscles and thrash about when they move.

Mollusca - clams, octopus; the soft-bodied animals. These animals have a muscular foot for movement. They breathe through gills and most are able to make a shell for protection from predators. They have an open circulatory system with sinuses bathing the body regions.

Annelida - segmented worms; the first with specialized tissue. The advanced circulatory system of these worms has blood vessels and is a closed system. The nephridia are their excretory organs. Annelida are hermaphroditic and each worm fertilizes the other upon mating. Segmented worms support themselves with a hydrostatic skeleton and have circular and longitudinal muscles for movement.

Arthropoda - insects, crustaceans, and spiders; this is the largest group of the animal kingdom. Phylum arthropoda accounts for about 85% of all the animal species. Arthropoda possess an exoskeleton made of chitin. They must molt to grow. They breathe through gills, trachea, or book lungs. Movement varies with members being able to swim, fly, and crawl. There is a division of labor among the appendages (legs, antennae, etc). This is an extremely successful phylum with members occupying diverse habitats.

Echinodermata - sea urchins and starfish; these animals have spiny skin. Their habitat is marine. They have tube feet for locomotion and feeding.

Chordata - all animals with a notocord or a backbone. The classes in this phylum include Agnatha (jawless fish), Chondrichthyes (cartilage fish), Osteichthyes (bony fish), Amphibia (frogs and toads; possess gills that are replaced by lungs during development), Reptilia (snakes, lizards; the first to lay eggs with a protective covering), Aves (birds; warm-blooded), and Mammalia (animals that do not lay eggs, possess mammary glands that produce milk, and are warm-blooded).

Sample Test Question and Rationale

90. **Which phylum accounts for 85% of all animal species?**
    (Easy)

    A. Nematoda
    B. Chordata
    C. Arthropoda
    D. Cnidaria

Answer: C. Arthropoda

The arthropoda phylum consists of insects, crustaceans, and spiders. They are the largest group in the animal kingdom.

Skill 4.3.2 Life functions and associated structures (digestion, excretion, nervous control, contractile systems and movement, support, integument, immunity, endocrine system)

Digestion

The function of the digestive system is to break food down into nutrients and absorb it into the blood stream where it can be delivered to all cells of the body for use in cellular respiration.

Essential nutrients are those nutrients that the body needs but cannot make. There are four groups of essential nutrients: essential amino acids, essential fatty acids, vitamins, and minerals.

There are ten essential amino acids humans need. A lack of these amino acids results in protein deficiency. There are only a few essential fatty acids.

Vitamins are organic molecules essential for a nutritionally adequate diet. Nutritionists have identified thirteen vitamins essential to humans.

There are two groups of vitamins: water-soluble (includes vitamin B complex and vitamin C) and water insoluble (vitamins A, D and K). Vitamin deficiencies can cause severe problems.

Unlike vitamins, minerals are inorganic molecules. Calcium is needed for bone construction and maintenance. Iron is important in cellular respiration and is a major component of hemoglobin.

Carbohydrates, fats, and proteins fuel the generation of ATP. Water is necessary to keep the body hydrated.

The teeth and saliva begin digestion by breaking down food into smaller pieces and lubricating it so it can be swallowed. The lips, cheeks, and tongue form a bolus or ball of food. It is carried down the pharynx by the process of peristalsis (wave-like contractions) and enters the stomach through the sphincter, which closes to keep food from going back up. In the stomach, pepsinogen and hydrochloric acid form pepsin, the enzyme that hydrolyzes proteins. Pepsin and other chemicals break down the food further and it is churned into acid chyme. The pyloric sphincter muscle opens to allow food to enter the small intestine.

Most nutrient absorption occurs in the small intestine. Its large surface area, a function of its length and protrusions called villi and microvilli, provides the primary absorptive surface into the bloodstream. After leaving the stomach, acidic chyme is neutralized in the small intestine to allow the enzymes necessary to breakdown food to function. Accessory organs such as the pancreas and liver produce these enzymes and bile. The liver makes bile, which breaks down and emulsifies fatty acids. Any remaining food then enters the large intestine. The large intestine functions to reabsorb water and produce vitamin K. The feces, or remaining waste, passes out through the anus.

Gastric ulcers are lesions in the stomach lining. Ulcers are mainly caused by bacteria, but are worsened by pepsin and acid.

Appendicitis refers to inflammation of the appendix. The appendix has no known function, is open to the intestine, and can be blocked by hardened stool or swollen tissue. A blocked appendix can cause bacterial infections and inflammation leading to appendicitis. If left untreated, the appendix can rupture allowing stool and infection to spill out into the abdomen. Without immediate surgery this condition can be life threatening. Symptoms of appendicitis include lower abdominal pain, nausea, loss of appetite and fever.

Circulation

The function of the closed circulatory system (**cardiovascular system**) is to carry oxygenated blood and nutrients to all cells of the body and return carbon dioxide waste to the lungs for expulsion. The heart, blood vessels, and blood make up the cardiovascular system. The structure of the heart is shown below:

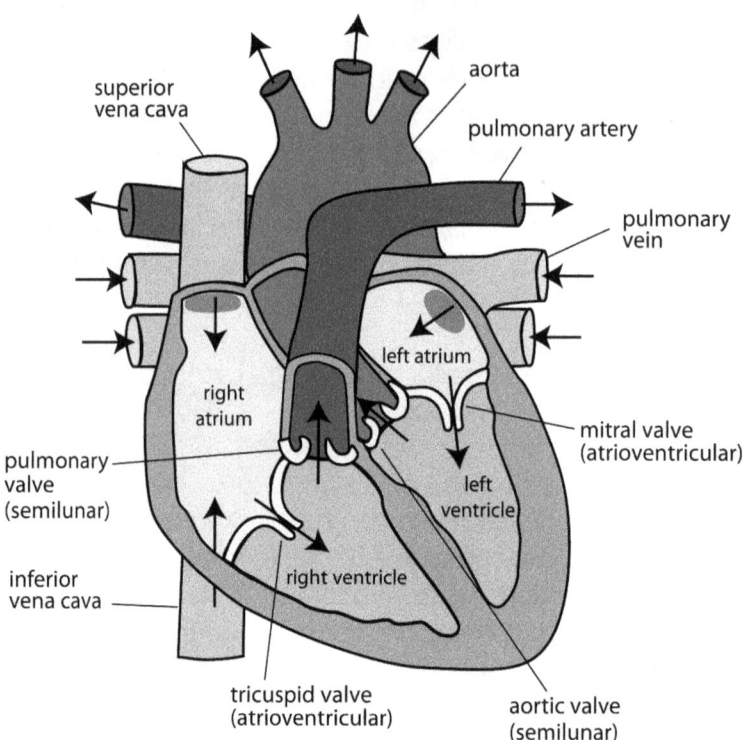

The atria are the chambers that receive blood returning to the heart while the ventricles are the chambers that pump blood out of the heart. There are four valves, two atrioventricular (AV) valves and two semilunar valves. The AV valves are located between each atrium and ventricle. The contraction of the ventricles closes the AV valve to keep blood from flowing back into the atria. The semilunar valves are located where the aorta leaves the left ventricle and where the pulmonary artery leaves the right ventricle. The semilunar valves are opened by ventricular contraction, allowing blood to be pumped out into the arteries, and are closed by ventricular relaxation.

Cardiac output is the volume of blood per minute that the left ventricle pumps. This output depends on heart rate and stroke volume. **Heart rate** is the number of times the heart beats per minute and **stroke volume** is the amount of blood pumped by the left ventricle per contraction. Humans have an average cardiac output of about 5.25 L/min. Heavy exercise can increase cardiac output up to five times. Epinephrine and increased body temperature also increase heart rate and cardiac output.

Cardiac muscle can contract without any signal from the nervous system. It is the sinoatrial node that is the pacemaker of the heart. The sinoatrial node is located on the wall of the right atrium and generates electrical impulses that make cardiac muscle cells contract in unison. The atrioventricular node shortly delays the electrical impulse to ensure the atria empty before the ventricles contract.

There are three kinds of blood vessels in the circulatory system: arteries, capillaries, and veins. **Arteries** carry oxygenated blood away from the heart to organs in the body. Arteries branch off to form smaller arterioles in organs. Arterioles form tiny **capillaries** that reach every tissue. Downstream, capillaries combine to form larger venules. Venules combine to form larger **veins** that return blood to the heart. Arteries and veins differ in the direction they carry blood.

Blood vessels are lined by endothelium. In veins and arteries, the endothelium is surrounded by a layer of smooth muscle and an outer layer of elastic connective tissue. Capillaries only consist of the thin endothelium layer and its basement membrane that allows for nutrient absorption.

Blood flow velocity decreases as it reaches the capillaries. The capillaries have the smallest diameter of all the blood vessels, but this is not why the velocity decreases. Arteries carry blood to such a large number of capillaries that the blood flow velocity actually decelerates as it enters the capillaries. Blood pressure is the hydrostatic force that blood exerts against the wall of a vessel. Blood pressure is greatest in arteries.

Blood is a connective tissue consisting of liquid plasma and several kinds of cells. Approximately 60% of the blood is plasma. Plasma contains water salts called electrolytes, nutrients, waste, and proteins. The electrolytes maintain a pH of about 7.4. The proteins contribute to blood viscosity and help maintain pH. Some of the proteins include clotting factors and immunoglobulins, the antibodies that help fend off infection.

The lymphatic system is responsible for returning lost fluid and proteins to the blood. Fluid enters lymph capillaries. This lymph fluid is filtered by lymph nodes that are filled with white blood cells to fight infection.

There are two classes of cells in blood, red blood cells and white blood cells. **Red blood cells (erythrocytes)** are the most numerous. They contain hemoglobin, which carries oxygen.

White blood cells (leukocytes) are larger than red blood cells. They are phagocytic and have the ability to engulf invaders. White blood cells are not confined to the blood vessels and can enter the interstitial fluid between cells. There are five types of white blood cells: monocytes, neutrophils, basophils, eosinophils, and lymphocytes.

A third cellular element found in blood is platelets. **Platelets** are made in the bone marrow and assist in blood clotting. The neurotransmitter that initiates blood vessel constriction following an injury is called serotonin. A material called prothrombin is converted to thrombin with the help of thrombokinase. The thrombin is then used to convert fibrinogen to fibrin, which traps red blood cells to form a scab and stop blood flow.

Cardiovascular diseases are the leading cause of death in the United States. Cardiac disease usually manifests as either a heart attack or stroke. During a heart attack, cardiac muscle tissue dies, usually from coronary artery blockage. During a stroke, nervous tissue in the brain dies due to the blockage of arteries in the head.

Many heart attacks and strokes are caused by a disease called atherosclerosis. Plaques form on the inner walls of arteries, narrowing the area in which blood can flow. Arteriosclerosis is when the arteries harden from plaque accumulation. Atherosclerosis can be prevented by a healthy diet that limits lipids and cholesterol and regular exercise. High blood pressure (hypertension) promotes atherosclerosis. Diet, medication, and exercise can reduce high blood pressure and prevent atherosclerosis.

Respiration

Animals constantly require oxygen for cellular respiration and need to remove carbon dioxide from their bodies. The respiratory surface must be large and moist. Different animal groups have different types of respiratory organs that perform gas exchange. Some animals use their entire outer skin for respiration (as in the case of worms). Fish and other aquatic animals have gills for gas exchange. Ventilation increases the flow of water over the gills. This process brings oxygen and removes carbon dioxide through the gills. Fish use a large amount of energy to ventilate its gills. This is because the oxygen available in water is less than that available in the air. Arthropoda (insects) have tracheal tubes that send air to all parts of their bodies.

Gas exchange for smaller insects is provided by diffusion. Larger insects ventilate their bodies using a series of body movements that compress and expand the tracheal tubes. All vertebrates, including humans, have lungs as their primary respiratory organ.

As the primary respiratory organ of the human respiratory system, the lungs contain a dense network of capillaries just beneath the epithelium. The surface area of the epithelium is about 100 m^2 in humans. Based on the surface area, the volume of air inhaled and exhaled is the tidal volume. This is normally about 500 mL in adults. Vital capacity is the maximum volume the lungs can inhale and exhale. This is usually around 3400 mL.

The respiratory system functions in the gas exchange of oxygen and carbon dioxide waste. It delivers oxygen to the bloodstream and picks up carbon dioxide for release out of the body. Air enters the mouth and nose, where it is warmed, moistened, and filtered of dust and particles. Cilia in the trachea trap unwanted material in mucus, which can be expelled. The trachea splits into two bronchial tubes and the bronchial tubes divide into smaller and smaller bronchioles in the lungs. The internal surface of the lung is composed of alveoli, which are thin walled air sacs. These allow for a large surface area for gas exchange. The alveoli are lined with capillaries. Oxygen diffuses into the bloodstream and carbon dioxide diffuses out of the capillaries to be exhaled out of the lungs. The oxygenated blood is carried to the heart and delivered to all parts of the body by hemoglobin, a protein consisting of iron.

The thoracic cavity holds the lungs. The diaphragm muscle below the lungs is an adaptation that makes inhalation possible. As the volume of the thoracic cavity increases, the diaphragm muscle flattens out and inhalation occurs.

Emphysema is a chronic obstructive pulmonary disease (COPD). Pulmonary diseases make it difficult for a person to breathe. Airflow through the bronchial tubes is partially blocked making breathing difficult. The primary cause of emphysema is cigarette smoke. People with a deficiency in alpha$_1$-antitrypsin protein production have a greater risk of developing emphysema at an earlier age. This protein helps protect the lungs from inflammatory damage. This deficiency is rare and can be tested for in individuals with a family history of disease. There is no cure for emphysema but there are treatments available. The best way to prevent emphysema is to avoid smoking.

Excretion

In many invertebrates osmoregulation and excretion involves tubular systems. The tubules branch throughout the body. Interstitial fluid enters these tubes and is collected into excretory ducts that empty into the external environment through openings in the body wall. Insects have excretory organs called Malpighian tubes. These organs pump water, salts and nitrogenous waste into the tubules. These fluids then pass through the hindgut and out the rectum.

Vertebrates have kidneys as the primary excretion organ. Both kidneys in the adult human are about 10 cm long. They receive about 20% of the blood pumped with each heartbeat despite their small size. The function of the excretory system is to rid the body of nitrogenous wastes in the form of urea.

The smallest functional unit of excretion in the kidney is the nephron. The structures of the kidney and the nephron are as follows:

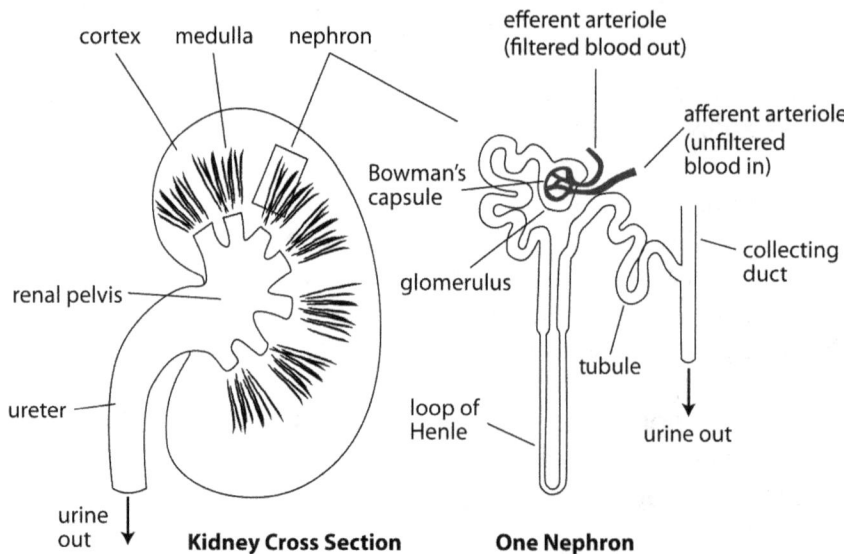

Kidney Cross Section   **One Nephron**

The Bowman's capsule contains the glomerulus, a tightly packed group of capillaries in the nephron. The glomerulus is under high pressure. Water, urea, salts and other fluids leak out due to pressure into the Bowman's capsule. This fluid waste (filtrate) passes through the three regions of the nephron: the proximal convoluted tubule, the loop of Henle, and the distal tubule. In the proximal convoluted tubule, unwanted molecules are secreted into the filtrate. In the loop of Henle, salt is actively pumped out of the tube and much water is lost due to the hyperosmosity of the inner part (medulla) of the kidney. As the fluid enters the distal tubule, more water is reabsorbed. Urine forms in the collecting duct that leads to the ureter then to the bladder where it is stored. Urine is passed from the bladder through the urethra. The amount of water reabsorbed back into the body is dependent upon how much water or fluids an individual has consumed. Urine can be very dilute or very concentrated depending on the level of concentration.

Glomerulonephritis (GN), often more generally called nephritis, usually occurs in children. Symptoms include hypertension, decreased renal function, hematuria and edema. Nephritis is produced by an antigen-antibody complex that causes inflammation and cell proliferation. Nephritis causes normal kidney tissue to become damaged and, if left untreated, can lead to kidney failure and death.

Nervous control

The central nervous system (CNS) consists of the brain and spinal cord. The CNS is responsible for the body's response to environmental stimuli. The spinal cord sends out motor commands in response to stimuli that are automated or reflexive. The brain is where responses to more complex stimuli occurs. The meninges are the connective tissues that protect the CNS. The CNS contains fluid filled spaces called ventricles. These ventricles are filled with cerebrospinal fluid which is formed in the brain. Cerebrospinal fluid cushions the brain and circulates nutrients, white blood cells, and hormones.

The **peripheral nervous system (PNS)** consists of the nerves that connect the CNS to the rest of the body. The sensory division brings information to the CNS from sensory receptors and the motor division sends signals from the CNS to effector cells. The motor division consists of the somatic nervous system and the autonomic nervous system. The somatic nervous system is controlled consciously in response to external stimuli. The autonomic nervous system is unconsciously controlled by the hypothalamus of the brain thereby regulating the body's internal environment. This system is responsible for the movement of smooth and cardiac muscles as well as the muscles for other organ systems.

The **neuron** is the basic unit of the nervous system. It consists of an axon, which carries impulses away from the cell body; the dendrite, which carries impulses toward the cell body; and the cell body, which contains the nucleus. The myelin sheath, comprised of Schwann cells, covers the neuron and provides insulation, which allow electrical impulses to travel quickly through the neuron. Synapses are junction between neurons. Chemicals called neurotransmitters serve as signaling molecules that are released from one neuron and diffuse through the synaptic cleft to another neuron.

Nerve action depends on an imbalance of electrical charges between the inside of the neuron and the outside. These electrical charges are carried by ions such as sodium, calcium, and potassium. When the ions move from one side of the neuronal membrane to the other (from outside the cell to inside or vice versa) an electrical current flows through the neuron. These electrical currents are called action potentials. Action potentials trigger the release of neurotransmitters from the axon into the synaptic cleft. When the neurotransmitters diffuse through the synaptic cleft, they bind to receptors on the surface of dendrites. This binding then triggers another action potential in the next neuron.

When a neuron is resting, it has a negative charge and is said to be hyperpolarized. When ions like sodium flow into the neuron it takes on a positive charge becoming depolarized and an action potential is generated.

Some neurons, or nerves, synapse on muscle cells. This is called a neuromuscular junction. In neuromuscular junctions there is a threshold of neurotransmitters that must be released by the nerve in order to generate a response from the muscle cell. This is called an "all or none" response.

There are many nervous system disorders. Parkinson's disease is caused by the degeneration of the basal ganglia, the brain region that controls motor movement. This causes a breakdown in the transmission of motor impulses to the muscles. Symptoms include tremors, slow movement, and muscle rigidity. Progression of Parkinson's disease occurs in five stages: early, mild, moderate, advanced, and severe. In the severe stage, an afflicted person is confined to a bed or chair. There is no cure for Parkinson's disease. Private research with stem cells is currently underway to find a cure for Parkinson's disease.

Contractile systems and movement

The function of the muscular system is to facilitate movement. There are three types of muscle tissue: skeletal, cardiac, and smooth.

Skeletal muscle is voluntary. These muscles are attached to bones and are responsible for their movement. Skeletal muscle consists of long fibers and is striated due to repeating patterns of myofilaments (made of the proteins actin and myosin) that make up the fibers.

Cardiac muscle is found in the heart. Cardiac muscle is striated like skeletal muscle, but differs in that the plasma membrane of the cardiac muscle causes the muscle to beat even when away from the heart. The action potentials of cardiac and skeletal muscles also differ.

Smooth muscle is involuntary. It is found in organs and enables functions such as digestion and respiration. Unlike skeletal and cardiac muscle, smooth muscle is not striated. Smooth muscle has less myosin and does not generate as much tension as the striated muscles.

The mechanism of skeletal muscle contraction involves a nerve impulse striking a muscle fiber. This causes calcium ions to flood the sarcomere. The myosin fibers creep along the actin, causing the muscle to contract. Once the nerve impulse has passed, calcium is pumped out and the contraction ends.
The axial skeleton consists of the skull and vertebral bones. The appendicular skeleton consists of the shoulder girdle, arms, legs and tailbones. Bone is a connective tissue. Parts of the bone include; (1) compact bone which provides strength, (2) spongy bone which contains red marrow to make blood cells, (3) yellow marrow in the center of long bones that stores fat cells, and (4) the periosteum which is the protective covering on the outside of the bone.

In addition to bones and muscles, ligaments and tendons are important joint components. A joint is a place where two bones meet. Joints enable movement. Ligaments attach bone to bone. Tendons attach bone to muscle. There are three types of joints:

1. Ball and socket – allows for rotational movement. There is a ball and socket joint between the shoulder and humerus. This type of joint allows the arms and legs to move in many different directions.
2. Hinge – movement is restricted to a single plane. An example of a hinge joint is between the humerus and ulna.
3. Pivot – allows for the rotation of the forearm at the elbow and the hands at the wrist.

Support

The support system provides structure for the body. There are various kinds of support systems in animals.

1. Lower invertebrates do not have a support system as such. Their bodies are made up of muscles and they either crawl or drag themselves about.

2. Soft bodied animals do not have a support structure, but almost all (excluding slugs) have shells. These shells offer protection from predators.

3. Complex invertebrates have exoskeletons, which are located outside the body. Exoskeletons are made up of a number of hardened plates, which support the animal, give it shape and protect it from predators.

4. Vertebrates have a support structure made up of a number of bead-like structures known as vertebrae. Vertebrates also have an endoskeleton. An endoskeleton is a skeleton that is present inside the body. While exoskeletons limit the growth of the body, endoskeletons do not.

A chordate is an animal that, at some time in its life, has a tough, flexible rod that runs along its back. Vertebrates are chordates. In most vertebrates, the rod along the back is replaced by a backbone.

Integument

The skin consists of two distinct layers. The epidermis is the thin outer layer and the dermis is the thick inner layer. Layers of tightly packed epithelial cells make up the epidermis. The tight packaging of the epithelial cells supports the skin's function as a protective barrier against outside elements.

The top layer of the epidermis consists of dead skin cells and is filled with keratin, a waterproofing protein. The dermis consists of connective tissue. It contains blood vessels, hair follicles, sweat glands, and sebaceous glands. An oily secretion called sebum, produced by the sebaceous gland, is released to the outer epidermis through the hair follicles. Sebum maintains the pH of the skin between 3 and 5, which inhibits most microbial growth.

The skin also plays a role in thermoregulation. Increased body temperature causes skin blood vessels to dilate, resulting in the loss of heat from the skin's surface. Sweat glands are also activated, increasing evaporative cooling. Decreased body temperature causes skin blood vessels to constrict. This results in the diversion of blood from the skin to deeper tissues thereby reducing heat loss from the surface of the skin.

Immunity

The immune system is responsible for defending the body against foreign invaders. There are two types of defense mechanisms: non-specific and specific.

The **non-specific** immune mechanism is comprised of two parts. The body's physical barriers are the first line of defense. These include the skin and mucous membranes. The skin prevents the penetration of bacteria and viruses as long as there are no abrasions on the skin. Mucous membranes form a protective barrier around the digestive, respiratory, and genitourinary tracts. In addition, the pH of the skin and mucous membranes inhibit the growth of many microbes. Mucous secretions (tears and saliva) wash away many microbes and contain lysozymes that kill many microbes.

The second component of the non-specific immune response includes white blood cells and inflammatory responses. **Phagocytosis** is the process of engulfing foreign particles with the cell membrane to form an internal phagosome. Neutrophils make up about seventy percent of all white blood cells. Monocytes mature to become macrophages, which are the largest phagocytic cells.

Eosinophils are also phagocytic. Natural killer cells destroy the body's own infected cells instead of the invading microbe directly. During an inflammatory response, blood supply to the injured area is increased, causing redness and heat. Swelling also typically occurs with inflammation. Histamine is released by basophils and mast cells when cells are injured triggering the inflammatory response.

The **specific** immune mechanism recognizes specific foreign material and responds by destroying the invader. These mechanisms are specific and diverse. They are able to recognize individual pathogens. An **antigen** is any foreign particle that elicits an immune response. An **antibody** is manufactured by the body to specifically recognize and latch onto antigens to destroy them. Memory of the invaders provides immunity upon further exposure.

Immunity is the body's ability to recognize and destroy an antigen before it causes harm. Active immunity develops after recovery from an infectious disease (e.g. chickenpox) or after a vaccination (e.g., mumps, measles, and rubella). Passive immunity may be passed from one individual to another and is not permanent. A good example is the immunity passed from mother to nursing child. A baby's immune system is not well developed and the passive immunity they receive through nursing provides them with additional protection.

There are two main responses made by the body after exposure to an antigen: humoral and cell-mediated.

1. **Humoral response** - Free antigens and antigen presenting cells activate B cells (lymphocytes from bone marrow) which transform into plasma cells that secrete antibodies. Memory cells are also generated that recognize future exposure to the same antigen. Antibodies defend the body against extracellular pathogens by binding to the antigen and making them an easy target for phagocytes to engulf and destroy. Antibodies are in a class of proteins called immunoglobulins. There are five major classes of immunoglobulins (Ig) involved in the humoral response: IgM, IgG, IgA, IgD, and IgE.

2. **Cell-mediated response** – Infected cells activate T cells (lymphocytes from the thymus) which then bind to the infected cells and destroy them along with the antigen. T cell receptors on T helper cells recognize antigens bound to the body's own cells. T helper cells release IL-2, which stimulates other lymphocytes (cytotoxic T cells and B cells). Cytotoxic T cells kill infected host cells by recognizing specific antigens.

Vaccines are antigens given in very small amounts. They stimulate both humoral and cell-mediated responses and help memory cells recognize future exposure to the antigen so antibodies can be produced much faster. The immune system attacks not only microbes, but also cells that are not native to the host such as skin grafts, organ transplantations, and blood transfusions. Antibodies to foreign blood and tissue types already exist in the body. If blood is transfused that is not compatible with the host, these antibodies destroy the new blood cells. There is a similar reaction when tissue and organs are transplanted.

The major histocompatibility complex (MHC) is responsible for the rejection of tissue and organ transplants. This complex is unique to each person. Cytotoxic T cells recognize the MHC on transplanted tissue or organ as foreign and destroy these tissues. Various drugs are needed to suppress the immune system and prevent rejection of foreign tissue; however, this also leaves the patient more susceptible to infection.

Autoimmune disease occurs when the body's own immune system destroys its own cells. Lupus, Grave's disease, and rheumatoid arthritis are examples of autoimmune diseases. There is no way to prevent autoimmune diseases. Immunodeficiency is a deficiency in either the humoral or cell-mediated immune defenses. HIV is an example of an immunodeficiency disease.

Endocrine system

The function of the **endocrine system** is to manufacture proteins called hormones. **Hormones** are released into the bloodstream and are carried to a target tissue where they stimulate an action. There are two classes of hormones: steroid and peptide. Steroid hormones come from cholesterol and include the sex hormones. Peptide hormones are derived from amino acids. Hormones are specific and fit receptors on the cell surface of the target tissue. Receptor binding then activates an enzyme that converts ATP to cyclic AMP. Cyclic AMP (cAMP) is a second messenger that travels from the cell membrane to the nucleus. When cAMP is triggered, the genes found in the nucleus turn on or off to cause a specific response.

Hormones are secreted by endocrine cells, which make up endocrine glands. The major endocrine glands and their hormones are as follows:

> **Hypothalamus** – located in the lower brain; signals the pituitary gland.
>
> **Pituitary gland** – located at the base of the hypothalamus; releases growth hormones and antidiuretic hormone (causing retention of water in kidneys).
>
> **Thyroid gland** – located on the trachea; lowers blood calcium levels (calcitonin) and maintains metabolic processes (thyroxine).
>
> **Gonads** – located in the testes of the male and the ovaries of the female; testes release androgens to support sperm formation and ovaries release estrogens to stimulate uterine lining growth and progesterone to promote uterine lining growth.
>
> **Pancreas** – secretes insulin to lower blood glucose levels and glucagon to raise blood glucose levels.

The thyroid gland produces hormones that help maintain heart rate, blood pressure, muscle tone, digestion, and reproductive functions. The parathyroid glands maintain the calcium level in blood and the pancreas maintains glucose homeostasis by secreting insulin and glucagon. The three gonadal steroids, androgen (testosterone), estrogen, and progesterone, regulate the development of the male and female reproductive organs.

Neurotransmitters are chemical messengers. The most common neurotransmitter is acetylcholine. Acetylcholine controls muscle contraction and heartbeat. A group of neurotransmitters, the catecholamines, include epinephrine and norepinephrine. Epinephrine (adrenaline) and norepinephrine are also hormones. They are produced in response to stress. They have profound effects on the cardiovascular and respiratory systems. These hormones/neurotransmitters can be used to increase the rate and stroke volume of the heart, thus increasing the rate of oxygen delivery to the blood cells.

Diabetes is the most well known endocrine disorder. Diabetes is caused by a deficiency of insulin that results in high blood glucose. Type I diabetes is an autoimmune disorder. The immune system attacks the cells of the pancreas, diminishing the ability to produce insulin. Treatment for type I diabetes consists of daily insulin injections. Type II diabetes usually occurs with age and/or obesity and involves a reduced response in target cells to insulin due to a reduction in the number of insulin receptors or a deficiency of insulin. Type II diabetics need to monitor their blood glucose levels. Treatment usually includes dietary restrictions and exercise.

Hyperthyroidism is another disorder of the endocrine system, resulting in excessive secretion of thyroid hormones. Symptoms are weight loss, high blood pressure, and high body temperature. The opposite condition, hypothyroidism, causes weight gain, lethargy, and intolerance to cold.

Sample Test Questions and Rationale

91. **Fats are broken down by which substance?**
    (Average Rigor)

    A. bile produced in the gall bladder
    B. lipase produced in the gall bladder
    C. glucagons produced in the liver
    D. bile produced in the liver

Answer: D. bile produced in the liver

The liver produces bile, which breaks down and emulsifies fatty acids.

TEACHER CERTIFICATION STUDY GUIDE

92. A boy had the chicken pox as a baby. He will most likely not get this disease again because of
    (Average Rigor)

    A. passive immunity
    B. vaccination.
    C. antibiotics.
    D. active immunity.

Answer: D. active immunity.

Active immunity develops after recovery from an infectious disease, such as the chicken pox, or after vaccination. Passive immunity may be passed from one individual to another (from mother to nursing child).

93. Movement is possible by the action of muscles pulling on
    (Average Rigor)

    A. skin.
    B. bones.
    C. joints.
    D. ligaments.

Answer: B. bones.

Skeletal muscles are attached to bones and are responsible for their movement.

94. Hormones are essential for the regulation of reproduction. What organ is responsible for the release of hormones for sexual maturity?
    (Average Rigor)

    A. pituitary gland
    B. hypothalamus
    C. pancreas
    D. thyroid gland

Answer: B. Hypothalamus

The hypothalamus begins secreting hormones that help mature the reproductive system and stimulate development of secondary sex characteristics.

TEACHER CERTIFICATION STUDY GUIDE

95. Which of the following compounds is not needed for skeletal muscle contraction to occur?
    (Rigorous)

    A. glucose
    B. sodium
    C. acetylcholine
    D. Adenosine 5'-triphosphate

Answer: A. Glucose

Although glucose is necessary to generate ATP (Adenosine 5'-triphosphate) it is not directly involved in muscular contractions. Acetylocholine is the neurotransmitter that intiates muscle contraction. Sodium plays an essential part in creating an action potential. Lastly, ATP provides the energy source for contraction.

96. Which of the following hormones is most involved in the process of osmoregulation?
    (Rigorous)

    A. Antidiuretic Hormone.
    B. Melatonin.
    C. Calcitonin.
    D. Gulcagon.

Answer: A. Antidiuretic Hormone.

The mechanism through which the body controls water concentration and various soluble materials is called osmoregulation. Antidiuretic Hormone (ADH) regulates the kidneys' reabsorption of water and directly affects the amount of water in the body. A failure to produce ADH can cause an individual to die from dehydration within a matter of hours. Calcitonin controls the removal of calcium from the blood. Glucagon, like insulin, controls the amount of glucose in the blood. Like ADH melatonin plays a role in homeostasis, by regulating body rhythms.

97. Capillaries come into contact with a large surface of both the kidneys and lungs, especially in relation to the volume of these organs. Which of the following is not consistent with both organs and their contact with capillaries.
(Rigorous)

   A. Small specialized sections of each organ contact capillaries
   B. A large branching system of tubes within the organ
   C. A large source of blood that is quickly divided into capillaries
   D. A sack that contains a capillary network

Answer: D. A sack that contains a capillary network

The Bowmen's capsule of the kidneys can be described as a sack that contains a capillary network. The alveoli of the lungs are sacks, however the capillaries are on the outside of the alveoli.

98. Which of the following substances in unlikely to cause negative consequences if over-ingested?
(Rigorous)

   A. essential fatty acids
   B. essential minerals
   C. essential water-insoluble vitamins
   D. essential water-soluble vitamins

Answer: D. essential water-soluble vitamins

Water-soluble vitamins are often removed in the filtration process by the kidneys and thus rarely build up to dangerous levels. Too many fatty acids can lead to obesity and other health problems. Excessive minerals can lead to a variety of different conditions, depending on the mineral ingested. Water-insoluble vitamins are usually stored in fatty tissues and thus are not flushed from the body. Therefore, water-insoluble vitamins can build up reaching dangerous levels.

99. If someone were experiencing unexplained changes in body temperature, hunger, and circadian rythyms, which of the following structures would most likely be the cause of these problems? *(Rigorous)*

    A. hypothalamus
    B. central nervous system
    C. pineal gland
    D. basal ganglia

Answer: A. hypothalamus

The pineal gland releases melatonin, which has been linked to sleep/wake patterns. The basal ganglia and central nervous system are structures regulating nerve impulses. Only the hypothalamus is responsible for regulating body temperature, hunger, and sleep/wake cycles.

Skill 4.3.3 Reproduction and development (gametogenesis, fertilization, parthenogenesis, embryogenesis, growth and differentiation, metamorphosis, aging)

Hormones regulate sexual maturation in humans. Humans cannot reproduce until puberty, about the age of 8-14, depending on the individual. The hypothalamus begins secreting hormones that stimulate maturation of the reproductive system and development of secondary sex characteristics. Reproductive maturity in girls occurs with their first menstruation and occurs in boys with the first ejaculation of viable sperm.

Hormones also regulate reproduction. In males, the primary sex hormones are the androgens, testosterone being the most important. The androgens are produced in the testes and are responsible for primary and secondary sex characteristics of the male. Female hormone patterns are cyclic and complex. Most women have a reproductive cycle length of about 28 days. The menstrual cycle is specific to the changes in the uterus. The ovarian cycle results in ovulation and occurs in parallel with the menstrual cycle. This parallelism is regulated by hormones. Five hormones participate in this regulation, most notably estrogen and progesterone. Estrogen and progesterone play an important role in signaling to the uterus and development and maintenance of the endometrium. Estrogens are also responsible for secondary sex characteristics of females.

Gametogenesis

Gametogenesis is the production of the sperm and egg cells.

Spermatogenesis begins at puberty in the male. One spermatogonia, the diploid precursor of sperm, produces four sperm. Immature sperm are produced in the seminiferous tubules located in the testes. After leaving the seminiferous tubules, sperm mature and are stored in the epididymis located on top of the testes. After ejaculation, the sperm travel up the **vas deferens** where they mix with semen made in the prostate and seminal vesicles and travel out the urethra.

Oogenesis, the production of egg cells (ova), is usually complete by birth in a female. Egg cells are not released until menstruation begins at puberty. Meiosis forms one ovum with all the cytoplasm and three polar bodies that are reabsorbed by the body. The ovum are stored in the ovaries and released each month from puberty to menopause.

Fertilization and embryogenesis

Ovulation releases the egg into ciliated fallopian tubes that move the egg along. Fertilization of the egg by the sperm normally occurs in the fallopian tube. If pregnancy does not occur, the egg passes through the uterus and is expelled through the vagina during menstruation. Levels of progesterone and estrogen stimulate menstruation and are affected by the implantation of a fertilized egg.

If fertilization occurs, the zygote begins dividing in approximately 24 hours. The resulting cells form a blastocyst that implants two to three days later in the uterus. Implantation promotes secretion of human chorionic gonadotrophin (HCG). This is what is detected in pregnancy tests. HCG keeps the level of progesterone elevated to maintain the uterine lining in order to feed the developing embryo until the umbilical cord forms.

Organogenesis, the development of the body organs, occurs during the first trimester of fetal development. The heart begins to beat and all the major structures are present at this time. The fetus grows very rapidly during the second trimester of pregnancy. The fetus is about 30 cm long and is very active during this stage. During the third and last trimester, fetal activity may decrease as the fetus grows. Labor is initiated by the hormone oxytocin, which causes dilation of the cervix and labor contractions. Prolactin and oxytocin cause the production of milk.

Parthenogenesis

Parthenogenesis is a type of asexual reproduction. Asexual reproduction, which does not require the union of egg and sperm to create a new organism, is common among bacteria, but unusual among animals. Parthenogenesis is when an embryo or seed grows and develops without fertilization. As with all asexual reproduction, there are both benefits and drawbacks to parthenogenesis. In the animal kingdom, parthenogenesis has been observed to occur naturally in invertebrates including water fleas, aphids, and honeybees and in vertebrates such as lizards, salamanders, certain fish, and turkeys.

Because there is no contribution from a male, the offspring of parthenogenetic embryos are female. However, the offspring themselves may be capable of sexual reproduction, parthenogenesis, or no reproduction at all.

Parthenogenesis does not simply produce clones. It is a true reproductive process and creates new individuals from the varied genetic material of the mother. For instance, a litter of kittens resulting from parthenogenesis would contain unique siblings created from the mother's genetic material, each from a different egg. Clearly, however, with all the genetic material coming from a single parent, there will be less overall genetic variability.

Certain species, such as the whiptail lizard (genus *Cnemidophorus*) reproduce exclusively via parthenogenesis. Other species may alternate between parthenogenesis and sexual reproduction. The alternation between parthenogenesis and sexual reproduction is called heterogamy. Sometimes, certain environmental stimuli trigger parthenogenesis. For example, aphids use parthenogenesis when there is ample food because parthenogenesis is more rapid than sexual reproduction. In some species, such as wasps, parthenogenesis is induced when they become infected with bacteria (genus *Wolbachia*).

Because *Wolbachia* pass to the new generation through eggs, but not through sperm, the bacteria have evolved to be able to force the wasps to reproduce through parthenogenesis. This is because it is advantageous for the bacterium that females be produced rather than males.

Natural parthenogenesis was first observed by Charles Bonnet in the 18th century. In 1900, Jacques Loeb became the first scientist to trigger artificial parthenogenesis using an unfertilized frog egg. Since that time, a large variety of mechanical and chemical stimuli have been used to achieve artificial parthenogenesis in most major groups of animals, although it does not always result in normal development of the offspring. Gregory Pincus's work with rabbits in 1936 showed that mammalian parthenogenesis was possible. Though it is widely believed to be possible, no successful attempts at human parthenogenesis have yet been reported. Some scientists believe that human parthenogenesis can provide a source of stem cells without the ethical dilemmas of embryonic stem cell use.

Growth and differentiation

Differentiation is the process in which cells become specialized in structure and function. The fate of the cell is usually maintained through many subsequent generations. Gene regulatory proteins can generate many cell types during development. Scientists believe that these proteins are passed down to the next generation of cells to ensure the specialized expression of genes.

Stem cells are not terminally differentiated. They can divide for as long as the animal is alive. When the stem cell divides, its daughter cells can either remain a stem cell or proceed with terminal differentiation. There are many types of stem cells that are specialized for different classes of terminally differentiated cells.

Embryonic stem cells give rise to all the tissues and cell types in the body. In culture, these cells have led to the creation of animal tissue that can replace damaged tissues. It is hopeful that with continued research, embryonic stem cells can be cultured to replace damaged muscles, tissues, and organs.

Animal tissue becomes specialized during development. The ectoderm (outer layer) becomes the epidermis or skin. The mesoderm (middle layer) becomes muscles and other organs beside the gut. The endoderm (inner layer) becomes the gut, also called the archenteron.

Development

Sponges are the simplest animals and lack true tissue. They exhibit no symmetry.

Diploblastic animals have only two germ layers: the ectoderm and endoderm. They have no true digestive system and exhibit radial symmetry. Diploblastic animals include the Cnideria (jellyfish).

Triploblastic animals have all three germ layers. Triploblastic animals can be further divided into: Acoelomates, Pseudocoelomates, and Coelomates.

> **Acoelomates** have no defined body cavity. An example is the flatworm (Platyhelminthe), which must absorb food from a host's digestive system.
>
> **Pseudocoelomates** have a body cavity that is not lined by tissue from the mesoderm. An example is the roundworm (Nematoda).

> **Coelomates** have a true fluid filled body cavity called a coelom derived from the mesoderm. Coelomates can further be divided into protostomes and deuterostomes. In the development of protostomes, the first opening becomes the mouth and the second opening becomes the anus. The mesoderm splits to form the coelom. In the development of deuterostomes, the mouth develops from the second opening and the anus from the first opening. The mesoderm hollows out to become the coelom. Protostomes include animals in the phyla Mollusca, Annelida, and Arthropoda. Deuterostomes include animals in phyla Ehinodermata and Vertebrata.

Development is defined as a change in form. Animals go through several stages of development after fertilization of the egg cell: cleavage, blastula, gastrulation, neuralation, and organogenesis.

> **Cleavage** - the first divisions of the fertilized egg. Cleavage continues until the egg becomes a blastula.

> **Blastula** - a hollow ball of undifferentiated cells.

> **Gastrulation** - the time of tissue differentiation into the separate germ layers, the endoderm, mesoderm, and ectoderm.

> **Neuralation** - development of the nervous system.

> **Organogenesis** - the development of the various organs of the body.

Metamorphosis

Metamorphosis is the process by which an animal progresses through relatively abrupt changes in physical form.

These changes in form are often accompanied by changes in behavior. The various stages of metamorphosis are known as instars. Metamorphosis occurs in insects, amphibians, mollusks, crustaceans, echinoderms, and tunicates. Insect metamorphosis has been extensively studied and is the focus of this section, however, the principles discussed below are applicable to all animals that metamorphose. Metamorphosis is usually classed as being either complete or incomplete.

Incomplete metamorphosis

About 12% of all insects experience this type of metamorphosis. Incomplete metamorphosis is also called **hemimetabolism**. It is common in insects in the order Mantodea, which includes the praying mantis. Incomplete metamorphosis consists of the following 3 stages:

1. **Egg** - The eggs are laid by the female and often are covered by an egg case to protect and hold them together.
2. **Nymph** - The eggs hatch and nymphs emerge. Though nymphs look similar to adults, they are smaller and typically lack wings. As nymphs, insects eat the same food that they will as adults. The nymphs shed and replace their exoskeletons in a process called molting or ecdysis. Most nymphs molt 4-8 times. In aquatic species, insects at this stage are called naiads.
3. **Adult** – The adult stage is primarily distinguished by the presence of wings. The insects will also stop molting when they reach their adult size.

Complete metamorphosis

The remaining 88% of insects go through complete metamorphosis, also known as **holometabolism**. Complete metamorphosis has the following 4 stages:

1. **Egg** – As above, the females lay eggs.
2. **Larva** – The larvae emerge from the eggs, but do not resemble the adult form of the insect. Typically, they have a worm-like shape. For example, caterpillars, maggots, and grubs are all larval stages of insects. Like the nymphs described above, larvae may molt several times as they grow larger.
3. **Pupa** – Pupa is also called chrysalis and describes the stage in which larvae make a cocoon around themselves. Inside the cocoon, larvae are unable to eat. During this phase, which may take anywhere from days to months, their bodies take on an adult shape including wings, legs, and internal organs.
4. **Adult** – After transformation from the pupa is complete, adult insects emerge from cocoons.

Within these two types of metamorphosis there are many variations. For instance, many species of beetles undergo hypermetamorphosis, which involves progression through several successive larval forms before pupation. There is similar variability in the amount of time various species spend in the stages detailed above. Extreme examples include the mayfly which has a non-eating adult stage that lives only one day and the cicada, which has a juvenile stage that lives underground for as long as 17 years.

Hormones produced by the endocrine glands control the growth of insects and the process of metamorphosis. Also, some cells within the insect's brain synthesize hormones that activate the thoracic gland, which in turn secretes a steroid, Ecdysone, which triggers metamorphosis. This same hormone also controls molting (ecdysis).

Aging

Aging or senescence is the combined effect of several processes of deterioration that typically follows the developmental stage of a living thing. The term senescence can refer to either cellular senescence or organismal senescence.

Cellular senescence

Cellular senescence is the loss of a cell's ability to divide. It is typically triggered by damage to the cell's DNA and often ends in intentional cell death (apoptosis) if the damage cannot be repaired. Apoptosis is still not fully understood, though some believe it evolved as a mechanism to prevent uncontrolled cell division (cancer). The more times a cell divides, the more chances it has to accumulate errors in its DNA. Therefore, a mechanism to prevent damaged cells from continuing to divide would seem logical. Scientists currently believe that telomeres (the highly repetitive DNA on the ends of chromosomes), which appear to shorten during each cell division, may play a regulatory role in limiting cell division and triggering senescence.

It is interesting to note that cellular senescence is not observed in all organisms. Single cell organisms and species including sponges, corals, and lobsters do not exhibit any evidence of cellular senescence. By studying cells from these animals, as well as other "immortal" cell lines, we may be able to learn more about the process of cellular senescence.

Organismal senescence

Organismal senescence typically describes the process of aging and is associated with an organism's declining ability to respond to stress, increasing risk of disease, and increasing difficulty maintaining balance in vital systems. As death is the final consequence of aging, differences in organisms' average life spans correlate to their "rate of aging". That is, differences in the rate of aging mean that a human being is old at 80 years, but a dog is elderly at 12 years. It should be noted that lifespan trends exist across species. For instance, in mammals, there is a correlation between increased species size and increased life expectancy, though there are exceptions.

Aging is affected by a variety of processes; the efficiency of DNA repair, the rate of free radical production, and the accumulation of unrepaired damage in cells, tissues, and organs. The precise mechanisms for declined efficiency of the various biochemical processes are not well understood. Some gerontologists believe aging can be slowed or even "cured" by using advanced molecular therapies to repair accumulated damage. Evidence showing that environmental changes have a significant effect on the life span of test subjects supports this theory. For example, caloric restriction significantly extends the lifespan of worms and monkeys.

Evolution and aging

Scientists continue to attempt to answer the question, "why do we age?" As above, some believe aging is just accumulated environmental damage. Others contend that errors amass within DNA during somatic cell division making aging inevitable.

TEACHER CERTIFICATION STUDY GUIDE

Other theories take an evolutionary view. Peter Medawar formalized the Mutation Accumulation theory of aging, which states: "The force of natural selection weakens with increasing age - even in a theoretically immortal population, provided only that it is exposed to real hazards of mortality. If a genetic disaster... happens late enough in individual life, its consequences may be completely unimportant." That is, diseases that affect an individual late in life will not prevent that individual from producing healthy offspring. Therefore, those diseases will continue to be passed along through the generations. Another theory, the Evolution Inbreed theory of aging, suggests that aging evolved to reduce the incidence of inbreeding since extensive inbreeding could reduce the overall success of a species. Finally, a theory hypothesized by George C. Williams involves pleiotropic genes, those that affect multiple traits. This theory states that if a gene gives young organisms an advantage it will accumulate in a species, even if it also causes difficulties for older individuals.

Sample Test Questions and Rationale

100. **Which is the correct sequence of embryonic development in a frog?**
    (Average Rigor)

    A. cleavage – blastula – gastrula
    B. cleavage – gastrula – blastula
    C. blastula – cleavage – gastrula
    D. gastrula – blastula – cleavage

Answer: A. cleavage – blastula – gastrula

Animals go through several stages of development after egg fertilization. The first step is cleavage which continues until the egg becomes a blastula. The blastula is a hollow ball of undifferentiated cells. Gastrulation follows and is the stage in which tissue differentiates into separate germ layers: the endoderm, mesoderm, and ectoderm.

101. **Fertilization in humans usually occurs in the:**
    (Easy)

    A. cervix.
    B. ovary.
    C. fallopian tubes.
    D. vagina.

Answer: C. fallopian tubes.

Fertilization of the egg by the sperm normally occurs in the fallopian tube. The fertilized egg then implants in the uterine lining for development.

BIOLOGY

102. Which of the following has no relation to female sexual maturity? *(Rigorous)*

   A. thyroxine
   B. estrogen
   C. testosterone
   D. luteinizing hormone

Answer: A. thyroxine

Thyroxine is a hormone associated with the regulation of the body's metabolism, and is not related to the maturation process. Luteinizng hormone stimulates cells in both the testes and ovaries. Estrogen is the hormone most frequently associated with female development. Studies have indicated that levels of testosterone increase as a female goes through puberty and drop off after she reaches her sexual peak.

Skill 4.3.4   Behavior (taxes, instincts, learned behaviors, communication)

Animal behavior is responsible for courtship leading to mating, communication between species, territoriality, aggression between animals, and dominance within a group. Animal communication is any behavior by one animal that affects the behavior of another animal. Animals use body language, sound, and smell to communicate. Perhaps the most common type of animal communication is the presentation or movement of distinctive body parts. Many species of animals reveal or conceal body parts to communicate with potential mates, predators, and prey. In addition, many species of animals communicate with sound. Examples of vocal communication include the mating "songs" of birds and frogs and warning cries of monkeys. Finally, many animals release scented chemicals to communicate with other animals. Pheromones are one class of scented chemicals that are important in reproduction and mating. Another class of distinctive odors, secreted from specialized glands, function to alert animals to the presence of others.

Innate behaviors are inborn or instinctual. An environmental stimulus such as the length of day or temperature results in a behavior. Hibernation among some animals is an innate behavior. **Learned behavior** is modified due to past experience.

DOMAIN 5.0 ECOLOGY

Competency 5.1 Populations (intraspecific competition, density factors, population growth, dispersion patterns, life-history patterns, social behavior)

A **population** is a group of individuals of one species that live in the same general area. Many factors affect population size and growth rate. Population size can depend on the total amount of life a habitat can support. This is the carrying capacity of the environment. Once the habitat runs out of food, water, shelter, or space, the carrying capacity decreases, and then stabilizes.

Competition is when two or more species in a community use the same resources. Competition is usually detrimental to both populations. Competition is often difficult to find in nature because competition between two populations is not continuous. Either the weaker population will cease to exist, or one population will evolve to utilize other available resources.

Limiting factors can affect population growth. As a population increases, competition for resources is more intense, and the growth rate declines. This is a **density-dependent** growth factor. The carrying capacity can be determined by the density-dependent factor. **Density-independent factors** affect individuals regardless of population size. The weather and climate are good examples. Temperatures that are too hot or cold may kill many individuals from a population that has not reached its carrying capacity.

A zero population growth rate occurs when the birth and death rates are equal in a population. Exponential growth occurs when there is an abundance of resources and the growth rate is at its maximum, called the intrinsic rate of increase. This relationship can be graphically represented in a growth curve. An exponentially growing population begins with little change and then rapidly increases as seen in the J-curve below.

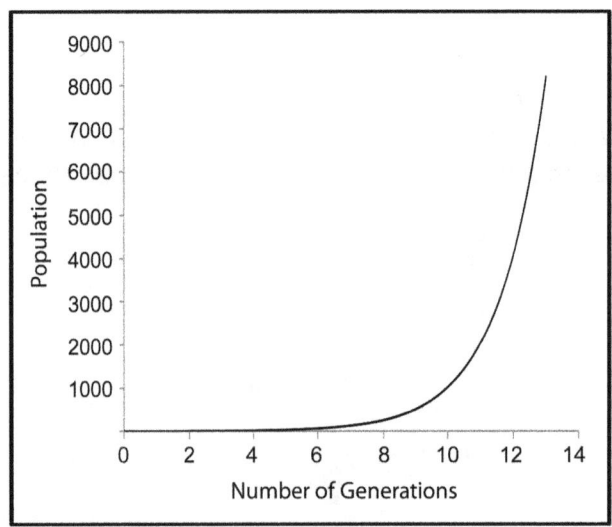

Logistic population growth incorporates the carrying capacity into the growth rate. As a population reaches the carrying capacity, the growth rate begins to slow down and level off as depicted in the S-curve below.

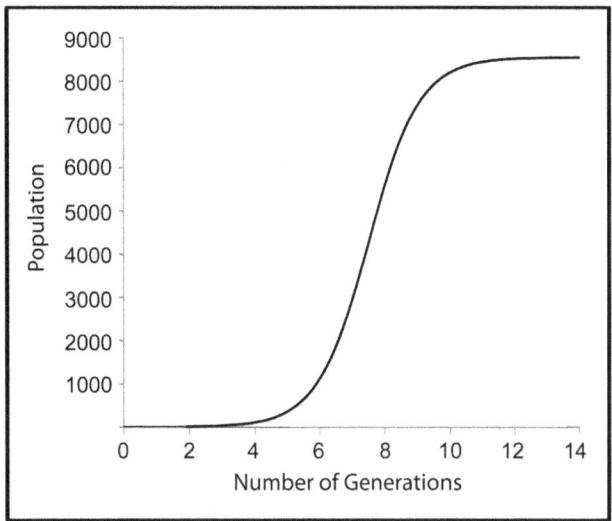

Many populations follow this model of population growth. Humans, however, are an exponentially growing population. Eventually, the carrying capacity of the Earth will be reached, and the growth rate will level off.

Population density is the number of individuals per unit area or volume. The spacing pattern of individuals in an area is called dispersion. **Dispersion patterns** can be clumped, with individuals grouped in patches; uniform, where individuals are approximately equidistant from each other; or random.

Population densities are usually estimated based on a few representative plots. Aggregation of a population in a relatively small geographic area can have detrimental effects on the environment. Rapid consumption of food, water, and other resources can result in an unstable environment. A low population density is less harmful to the environment. The use of natural resources will be more widespread, allowing adequate time for the environment to recover.

Life-history patterns

In the study of ecological populations, life history patterns are the sum total of traits that affect the reproduction and survival of an organism. The Darwinian goal of species evolution is to maximize reproductive output. Thus, different species vary in different parameters such as clutch size (number of offspring per reproductive event), frequency of reproduction, and investment in parental care to best balance reproductive output with individual survival. As is the case with most characteristics, a particular life-history pattern develops over time through natural selection. Natural selection tends to maximize the representation of those individuals that possess the best combination of life-history patterns (e.g., rate of offspring production, length of reproductive life span) that serve to promote high reproductive output.

Sample Test Questions and Rationale

103. **In the growth of a population, initially the increase is exponential until carrying capacity is reached. This is represented by a(n):**
(Average Rigor)

    A. S curve.
    B. J curve.
    C. M curve.
    D. L curve.

Answer: A. S curve.

An exponentially growing population starts off with little change and then rapidly increases. The graphic representation of this growth curve has the appearance of a "J". However, as the carrying capacity of the growing population is reached, the growth rate begins to slow down and level off. The graphic representation of this growth curve has the appearance of an "S".

104. **All of the following are density dependent factors that affect a population except**
    (Rigorous)

    A. disease.
    B. drought.
    C. predation.
    D. migration.

Answer: B. drought.

Although drought would affect the amount of food available to a population (which creates a density dependent factor), the drought itself would occur regardless of population size, and is thus density independent. Disease and migration tend to occur more frequently in crowded populations. The amount of prey and predators would affect the number of individuals in a population.

105. **Which of the following is not an example of dynamic equilibrium?**
    (Rigorous)

    A. a stable population
    B. a symbiotic pair of organisms
    C. osmoregulation
    D. maintaining head position while walking

Answer: D. maintaining head position while walking

Maintinaing head position while walking is a case of static equilibrium, a state where things do not change. In a stable population birth and death rates must balance. In a symbiotic pair, the contributions of each organism must balance, or the relationship becomes parasitic. Osmoregulation balances the body's need for water and the dissolved compounds within it.

106. All of the following are density independent factors that affect a population except
   (Average Rigor)

   A. temperature.
   B. rainfall.
   C. predation.
   D. soil nutrients.

Answer: C. predation

As a population increases, the competition for resources intensifies and the growth rate declines. This is a density-dependent factor. An example of this would be predation. Density-independent factors affect the population regardless of its size. Examples of density-independent factors are rainfall, temperature, and soil nutrients.

Competency 5.2   Communities (niche, interspecific relationships, species diversity, succession)

Niche

The term 'niche' describes the relational position of a species or population in an ecosystem. Niche includes how a population responds to the relative abundance of its resources and enemies (e.g., by growing when resources are abundant and predators, parasites, and pathogens are scarce). Niche also indicates the life history of an organism, habitat, and place in the food chain. According to the competitive exclusion principle, no two species can occupy the same niche in the same environment for a long time.

The full range of environmental conditions (biological and physical) under which an organism can exist describes its fundamental niche. Because of the pressure from superior competitors, organisms are driven to occupy a niche much narrower than their previous niche. This is known as the 'realized niche.'

Examples of niche:

1. Oak trees:

* live in forests
* absorb sunlight by photosynthesis
* provide shelter for many animals
* act as support for creeping plants
* serve as a source of food for animals
* cover the ground with dead leaves in the autumn

If the oak trees were cut down or destroyed by fire or storms they would no longer be doing their job and this would have a disastrous effect on all the other organisms living in the same habitat.

2. Hedgehogs:

* eat a variety of insects and other invertebrates, which live underneath dead leaves and twigs in the garden
* their spines are a superb environment for fleas and ticks
* put the nitrogen back into the soil when they urinate
* eat slugs thereby protecting plants from eating them

If hedgehogs ceased to exist, the slug population would explode and the nutrients in the dead leaves and twigs would not be recycled.

Interspecific relationships

Predation and **parasitism** are beneficial for one species and detrimental for the other. Predation is when a predator eats its prey. The common conception of predation is of a carnivore consuming other animals. This is one form of predation. Although not always resulting in the death of the plant, herbivory is also a form of predation. Some animals eat enough of a plant to cause death. Parasitism involves a predator that lives on or in its host, causing detrimental effects to the host. Insects and viruses living off of and reproducing in their hosts is an example of parasitism. Many plants and animals have evolved defenses against predators. Some plants have poisonous chemicals that will harm the predator if ingested and some animals are camouflaged so they are more difficult to detect.

Symbiosis is when two species live close together. Parasitism is one example of a symbiotic relationship. Another example of symbiosis is commensalism. **Commensalism** occurs when one species benefits from the other without causing any harm to the other species. **Mutualism** is when both species benefit from one another. Species involved in mutualistic relationships must co-evolve to survive. As one species evolves, the other must as well if it is to be successful in life. The grouper fish and a species of shrimp live in a mutualistic relationship. The shrimp feed off parasites living on the grouper. Thus, the shrimp are fed and the grouper stays healthy. Many microorganisms exist in mutualistic relationships.

Species diversity

Species diversity is simply a count of the number of different species in a given area. A species is a group of plants or animals that are similar, able to breed, and produce viable offspring.

Biologists are unsure of how many different species live on the earth. Estimates range from 2 - 100 million. So far, only 2.1 million species have been classified. Most of these species live in the middle latitudes. Most of the species that remain unclassified are invertebrates. This group includes insects, spiders, worms, and crustaceans. It is difficult to classify invertebrates because of their small size and the inaccessibility of the habitats in which they live. Another habitat that is relatively inaccessible is the tropical rain forest. It is estimated that this single biome may contain 50 - 90% of the earth's biodiversity.

Many species have become extinct over the course of geological history. Driving factors for extinction include extreme fluctuations in the environment and increased competition from superior species. Because of the industrial revolution, a large number of biologically classified species have become extinct. The continued extinction of species caused by human activities is one of the greatest environmental problems we currently face.

Species diversity is one of the three categories of biodiversity. The other two are genetic diversity (refers to the total number of genetic characteristics expressed and recessed in all of the individuals that comprise a particular species) and ecosystem diversity (variation of habitats, community types and non-living chemical and physical components in a given area).

Species richness is an important component of an ecosystem's biodiversity. Species richness is measurable in practice and has been found to be a good substitute for other measures of biodiversity that are difficult to measure.

A few facts about species richness:

1. Species richness is a measure of biodiversity.
2. Species richness increases from high latitudes to low latitudes.
3. Maximal species richness occurs between 20 - 30° N latitude, which includes the tropics region.
4. Larger areas contain more species, since there are more opportunities for species to live there.
5. Generally, the relationship between species richness and a species being endemic to a specific area are positively correlated. However, there are some islands in which there is a high degree of endemism, but a low level of species richness.

Succession

Succession is an orderly process of replacing a community that has been damaged or has previously ceased to exist. Primary succession occurs where life never existed before, as in a flooded area or a new volcanic island. Secondary succession takes place in communities that were once flourishing but were disturbed by some source, either man or nature, but not totally stripped. A climax community is a community that is established and flourishing.

Abiotic and biotic factors play a role in succession. **Biotic factors** are living things in an ecosystem (e.g., plants, animals, bacteria, and fungi). **Abiotic factors** are non-living aspects of an ecosystem (e.g., soil quality, rainfall, and temperature).

Abiotic factors affect succession by impacting which species can colonize an area. Certain species will or will not survive depending on the weather, climate, or soil makeup. Biotic factors such as inhibition of one species due to another may occur. This may be due to some form of competition between the species.

Sample Test Questions and Rationale

107. **If the niches of two species overlap, what usually results?**
 (Easy)

 A. a symbiotic relationship
 B. cooperation
 C. competition
 D. a new species

Answer: C. competition

Two species that occupy the same habitat or eat the same food are said to be in competition with one another.

108. **Primary succession occurs after...**
 (Average Rigor)

 A. nutrient enrichment.
 B. a forest fire.
 C. exposure of a bare rock after the water table permanently recedes.
 D. a housing development is built.

Answer: C. exposure of a bare rock after the water table permanently recedes.

Primary succession occurs where life never existed before, such as flooded areas or a new volcanic island. It is only after the water recedes that the rock is able to support new life.

109. A clownfish is protected by the sea anemone's tentacles. In turn, the anemone receives uneaten food from the clownfish. This is an example of
(Easy)

   A. mutualism.
   B. parasitism.
   C. commensalism.
   D. competition.

Answer: A. mutualism.

Neither the clownfish nor the anemone cause harmful effects towards one another and they both benefit from their relationship. Mutualism is when two species that occupy a similar space benefit from their relationship.

110. Which of the following are reasons to maintain biological diversity?

   I. Consumer product development.
   II. Stability of the environment.
   III. Habitability of our planet.
   IV. Cultural diversity.
   (Rigorous)

   A. I and III
   B. II and III
   C. I, II, and III
   D. I, II, III, and IV

Answer: D. I, II, III, and IV

Biological diversity refers to the extraordinary variety of living things and ecological communities throughout the world. Maintaining biological diversity is important for many reasons. First, we derive many consumer products from living organisms in nature. Second, the stability and habitability of the environment depends on the varied contributions of many different organisms. Finally, the cultural traditions of human populations depend on the diversity of the natural world. The answer is (D).

111. Which of the following is not an abiotic factor?
    (Easy)

    A. temperature
    B. rainfall
    C. soil quality
    D. bacteria

Answer: D. bacteria

Abiotic factors are the non-living aspects of an ecosystem. Bacteria is an example of a biotic factor, a living thing.

Competency 5.3   Ecosystems (terrestrial, aquatic, biomes, energy flow, biogeochemical cycles, stability and disturbances, human impact, interrelationships among ecosystems)

An ecosystem includes all living things in a specific area and the non-living things that affect them. The term **biome** is usually used to classify the general types of ecosystems in the world. Each biome contains distinct organisms best adapted to that natural environment (including geological make-up, latitude, and altitude). All the living things in a biome are naturally in equilibrium and disturbances in any one element may cause upset throughout the system. It should be noted that these biomes may have various names in different areas. For example steppe, savanna, veld, prairie, outback, and scrub are all regional terms that describe the same biome, a grassland.

The major terrestrial biomes are desert, grassland, tundra, boreal forest, tropical rainforest, and temperate forest.

Desert

Deserts refer to any location that receives less that 50 cm of precipitation a year. Despite their lack of water and often desolate appearance, the soils, though loose and silty, tend to be rich. Specialized plants and animals populate deserts. Plant species include xerophytes and succulents. Animals tend to be non-mammalian and small (e.g., lizards and snakes). Large animals are not able to find sufficient shade in the desert and mammals, in general, are not well adapted to store water and withstand heat. Deserts are dry and may be either cold or hot. Hot and dry deserts are what we typically envision when we think of a desert. Hot deserts are located in northern Africa, southwestern United States, and the Middle East. Similarly, cold deserts have little vegetation, small animals, and are located exclusively in Antarctica, Greenland, much of central Asia, and the Arctic. Both types of deserts do receive precipitation in the winter, though it is in the form of rain in hot deserts and the form of snow in the cold.

Grassland

As the name suggests, grasslands include large expanses of grass with only a few shrubs or trees. There are both tropical and temperate grasslands.

Tropical grasslands cover much of Australia, South America and India. The weather is warm year-round with moderate rainfall. However, the rainfall is concentrated in half the year and drought and fires are common in the other half of the year. These fires serve to renew rather than destroy areas within tropical grasslands. This type of grassland supports a large variety of animals from insects to mammals such as squirrels, mice, gophers, giraffes, zebras, kangaroos, lions, and elephants.

Temperate grasslands receive less rain than tropical grasslands and are found in South Africa, Eastern Europe and the western United States. As in tropical grasslands, periods of drought and fire serve to renew the ecosystem. Differences in temperature also differentiate the temperate from the tropical grasslands. Temperate grasslands are cooler in general and experience even colder temperatures in winter. These grasslands support similar types of animals as the tropical grasslands: prairie dogs, deer, mice, coyotes, hawks, snakes, and foxes.

The savanna is grassland with scattered individual trees. Plants of the savanna include shrubs and grasses. Temperatures range from 0 - 25° C in the savanna depending on its location. Rainfall is from 90 to 150 cm per year. The savanna is a transitional biome between the rain forest and the desert that is located in central South America, southern Africa, and parts of Australia.

Tundra

Tundras are treeless plains with extremely low temperatures (-28 to 15° C) and little vegetation or precipitation. Rainfall is limited, ranging from 10 to 15 cm per year. A layer of permanently frozen subsoil, called permafrost, is found in the tundra. Permafrost means that no vegetation with deep root systems can exist in the tundra, but low shrubs, mosses, grasses, and lichen are able to survive. These plants grow low and close together to resist the cold temperature and strong winds. The few animals that live in the tundra are adapted to the cold winters (via layers of subcutaneous fat, hibernation, or migration) and raise their young quickly during the summer. Such species include lemmings, caribou, arctic squirrels and foxes, polar bears, cod, salmon, mosquitoes, falcons, and snow birds.

Both arctic and alpine tundra exist, though their characteristics are extremely similar and are distinguished mainly by location (arctic tundra is located near the north pole, while alpine tundra is found in the world's highest mountains).

Polar tundra or permafrost temperature ranges from -40 to 0° C. It rarely gets above freezing. Rainfall is below 10 cm per year. Most of the water is found as ice and there are few living organisms.

Forests

There are three types of forests, all characterized by the abundant growth of trees, but with different climates, flora and fauna.

Boreal forest (taiga)

These forests are located throughout northern Europe, Asia, and North America, near the poles. The climate typically consists of short, rainy summers followed by long, cold winters with snow. The trees in boreal forests are adapted to cold winters and are typically evergreens including pine, fir and spruce. The trees are so thick that there is little undergrowth. A number of animals are adapted to life in the boreal forest, including many mammals such as bear, moose, wolves, chipmunks, weasels, mink and deer. These coniferous forests have temperatures ranging from -24 to 22° C. Rainfall is between 35 to 40 cm per year. This is the largest terrestrial biome.

Tropical rainforest

Tropical rainforests are located near the equator and are typically warm and wet throughout the entire year. The temperature is constant (25° C) and the length of daylight is about 12 hours. Precipitation is frequent and occurs evenly throughout the year. In a tropical rainforest, rainfall exceeds 200 cm per year. Tropical rainforests have abundant, diverse species of plants and animals. A tropical dry forest gets scarce rainfall and a tropical deciduous forest has wet and dry seasons. The soil is surprisingly nutrient-poor and most of the biomass is located within the trees themselves. Vegetation is highly diverse including many trees with shallow roots, orchids, vines, ferns, mosses and palms. Similarly, animals are plentiful and include all type of birds, reptiles, bats, insects and small to medium sized mammals.

Temperate forest

These forests have well defined winters and summers with precipitation throughout the year. Temperate forests are common in western Europe, eastern North America, and parts of Asia. Common trees include deciduous species such as oak, beech, maple and hickory. Unlike the boreal forests, the canopy in temperate forests is not particularly heavy so various smaller plants occupy the understory. Mammals and birds are the predominate form of animal life. Typical species include squirrels, rabbits, skunks, deer, bobcats, and bear. The temperature here ranges from -24 to 38° C. Rainfall is between 65 and 150 cm per year.

A subtype of temperate forests is Chaparral forests. Chaparral forests experience mild, rainy winters and hot, dry summers. Trees do not grow as well here. Spiny shrubs dominate. Regions of chaparral forests include the Mediterranean, California coastline, and southwestern Australia.

Aquatic ecosystems

Aquatic ecosystems are, as the name suggests, ecosystems located within bodies of water. Aquatic biomes are divided between fresh water and marine systems. Freshwater ecosystems are closely linked to terrestrial biomes. Lakes, ponds, rivers, streams, and swamplands are examples of freshwater biomes. Marine areas cover 75% of the earth. This biome is organized by the depth of the water. The intertidal zone is from the tide line to the edge of the water. The littoral zone is from the water's edge to the open sea. It includes coral reef habitats and is the most densely populated area of the marine biome. The open sea zone is divided into the epipelagic zone and the pelagic zone. The epipelagic zone receives more sunlight and has a larger number of species. The ocean floor is called the benthic zone and is populated with bottom feeders. Marine biomes include coral reefs, estuaries, and several other systems within the oceans.

Oceans

Within the world's oceans, there are several separate zones, each with its own temperature profiles and unique species. These zones include intertidal, pelagic, benthic, and abyssal. The intertidal and pelagic zones are further distinguished by the latitude at which they occur (species have evolved to live in various water temperatures). The intertidal zone is the shore area, which is alternately under and above the water, depending on the tides. Algae, mollusks, snails, crabs, and seaweed are all found in intertidal zones. The pelagic zone is further from land but near the ocean surface. This zone is sometimes called the euphotic zone. Temperatures are much cooler here than in the intertidal zone and organisms in this zone include surface seaweeds, plankton, various fish, whales and dolphins. Further below the ocean's surface is the benthic zone, which is even colder and darker. A lot of seaweed is found in this zone, as well as, bacteria, fungi, sponges, anemones, sea stars, and some fish. Deeper still is the abyssal zone, which is the coldest and darkest area of the ocean. The abyssal zone has a high pressure and low oxygen content. Thermal vents found in the abyssal zone support chemosynthetic bacteria, which are in turn consumed by invertebrates and fish.

Coral reefs

Coral reefs are located in warm, shallow water near large land masses. The most well known example is the Great Barrier Reef off the coast of Australia. The coral itself is the dominant life form in the reefs and obtains its nutrients largely through photosynthesis (performed by the algae). Many other animal life forms also populate coral reefs such as fish, octopuses, sea stars, and urchins.
Estuaries

Estuaries are found where fresh and sea water meet, for instance where rivers flow into the oceans. Many species have evolved to thrive in the unique salt concentrations that exist in estuaries. These species include marsh grasses, mangrove trees, oysters, crabs, and certain waterfowl.

Ponds and Lakes

As with other aquatic biomes, varied ecosystems are found in ponds and lakes. This is not surprising since lakes vary in size and location. Some lakes are even seasonal, lasting only a few months each year. In addition, within lakes there are zones, comparable to those in oceans. The littoral zone, located near the shore and at the top of the lake, is the warmest and lightest zone. Organisms in this zone typically include aquatic plants and insects, snails, clams, fish, and amphibians.

Further from land, but still at the surface of the lake is the limnetic zone. Plankton are abundant in the limnetic zone and are at the bottom of the food chain in this zone, ultimately supporting all freshwater fish. Deeper in the lake is the profundal zone, which is cooler and darker. As plankton die they fall to the bottom of the lake and therefore also serve as a valuable food source in the profundal zone. Again, small fish eat the dead plankton and are the basis for supporting life in this zone.

Rivers and Streams

This biome includes moving bodies of water. As expected, the organisms found within streams vary according to latitude and geological features. Additionally, characteristics of the stream change as it flows from its headwaters to the sea. Furthermore, as the depth of the river increases, zones similar to those seen in the ocean are seen. That is, different species live in the upper, sunlit areas (e.g., algae, top feeding fish, and aquatic insects) and in the darker, bottom areas (e.g., catfish, carp, and microbes).

Wetlands

Wetlands are the only aquatic biome that is partially land-based. They are areas of standing water in which aquatic plants grow. These species, called hydrophytes are adapted to extremely humid and moist conditions and include lilies, cattails, sedges, cypress, and black spruce. Animal life in wetlands includes insects, amphibians, reptiles, many birds, and a few small mammals. Though wetlands are usually classified as a freshwater biome, they are in fact salt marshes that support shrimp, various fish, and grasses.

Energy flow

Trophic levels are based on the feeding relationships that determine energy flow and chemical cycling. Autotrophs are the primary producers of the ecosystem. **Producers** mainly consist of plants. **Primary consumers** are the next trophic level. The primary consumers are the herbivores that eat plants or algae. **Secondary consumers** are the carnivores that eat the primary consumers. **Tertiary consumers** eat the secondary consumer. These trophic levels may go higher depending on the ecosystem. **Decomposers** are consumers that feed off animal waste and dead organisms. This pathway of food transfer is depicted by a food chain.

Most food chains are more elaborate, becoming food webs. Energy is lost as the trophic levels progress from producer to tertiary consumer. The amount of energy that is transferred between trophic levels is called ecological efficiency. The visual of this energy flow is represented in a **pyramid of productivity**.

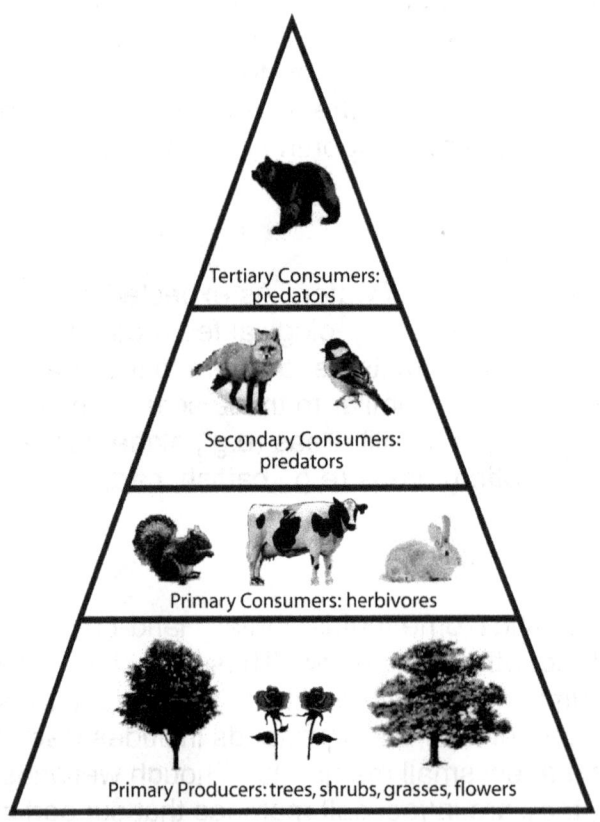

Depicted above, a **biomass pyramid** represents the total dry weight of organisms in each trophic level. A **pyramid of numbers** is a representation of the population size of each trophic level. The producers, being the most populous, are on the bottom of this pyramid with the tertiary consumers on the top with the fewest numbers.

Biogeochemical cycles

Biogeochemical cycles are nutrient cycles that involve both biotic and abiotic factors.

Water cycle - Two percent of all the available water is fixed and unavailable in ice or the bodies of organisms. Available water includes surface water (e.g., lakes, oceans, rivers) and ground water (e.g., aquifers, wells). Ninety six percent (96%) of all available water is ground water. The water cycle is driven by solar energy. Water is recycled through the processes of evaporation and precipitation. The water present now is the water that has been here since our atmosphere formed.

Carbon cycle - Ten percent (10%) of all available carbon in the air (in the form of carbon dioxide gas) is fixed by photosynthesis. Plants fix carbon in the form of glucose. Animals eat the plants and are able to obtain carbon. When animals release carbon dioxide through respiration, the plants again have a source of carbon for further fixation.

Nitrogen cycle - Eighty percent (80%) of the atmosphere is in the form of nitrogen gas. Nitrogen must be fixed and taken out of the gaseous form to be incorporated into an organism. Only a few genera of bacteria have the correct enzymes to break the triple bond between nitrogen atoms in a process called nitrogen fixation. These bacteria live within the roots of legumes (e.g., peas, beans, alfalfa) and add nitrogen to the soil so it may be absorbed by the plant. Nitrogen is necessary to make amino acids and the nitrogenous bases of DNA.

Phosphorus cycle - Phosphorus exists as a mineral and is not found in the atmosphere. A mutualistic symbioses between fungi (Myco) and plant roots (rhiza) called mycorrhizae fix insoluble phosphates into useable phosphorus. Urine and decayed matter return phosphorus to the earth where it can be fixed in plants. Phosphorus is required for the manufacture of ATP and DNA.

Role of Decomposers

Decomposers recycle the carbon accumulated in durable organic material that does not immediately proceed to the carbon cycle. Ammonification is the decomposition of organic nitrogen back into ammonia. This process of the nitrogen cycle is carried out by aerobic and anaerobic bacterial and fungal decomposers. Decomposers add phosphorous back to the soil by decomposing the excretion of animals.

Stability and disturbances

Nature replenishes itself continually. Natural disturbances such as landslides and brushfires are not solely destructive, they allow for a new generation of organisms to inhabit the land. For every indigenous organism, there exists a natural predator. These predator/prey relationships allow populations to maintain in reproductive balance and prevent over-consumption of food sources, thus keeping food chains in check. Left alone, nature always finds a way to balance itself. Unfortunately, the largest disturbances nature faces are created by humans. For example, humans have introduced non-indigenous species to many areas upsetting predator/prey relationships, building construction has caused landslides and disrupted waterfront ecosystems, and over utilization of land has depleted the ozone layer.

Human impact

The human population has been growing exponentially for centuries. People are living longer and healthier lives than ever before. Better health care and nutrition practices have contributed to enhanced human survival.

Human activity affects parts of nutrient cycles by removing nutrients from one part of the biosphere and adding them to another. This results in nutrient depletion in one area and nutrient excess in another. This affects water systems, crops, wildlife, and humans.

Humans are responsible for the depletion of the ozone layer. This depletion is due to the chemicals used for refrigeration and aerosols. The consequences of ozone depletion will be severe. The ozone protects the Earth from the majority of UV radiation. An increase in UV radiation will increase the incidence of skin cancer and have numerous other unknown effects on wildlife and plants.

Humans have a tremendous impact on the world's natural resources. The world's natural water supplies are affected by human use. Waterways are major sources for recreation and freight transportation. Oil and waste from boats and cargo ships pollute the aquatic environment adversely impacting aquatic animal and plant life.

Deforestation for urban development has resulted in the extinction or relocation of several species of plants and animals. Animals are forced to leave their forest homes or perish amongst the destruction. The number of plant and animal species that have become extinct due to deforestation is unknown. Scientists have only identified a fraction of the species on Earth.

Humans are continuously searching for new places to form communities. This encroachment on the environment leads to the destruction of wildlife communities. Conservationists focus on endangered species, but the primary focus should be on protection of the entire biome. If a biome becomes extinct, wildlife will die or invade another biome. Preservations established by the government aim at protecting small parts of biomes. While beneficial in the conservation of a few areas, the majority of the environment is still unprotected.

TEACHER CERTIFICATION STUDY GUIDE

Interrelationships among ecosystems

An ecosystem is the collection of all components and processes that define a portion of the biosphere. Ecosystems include both biotic (living) and abiotic (non-living) components. While individual organisms in an ecosystem affect other members of the ecosystem, ecosystems themselves are also interrelated. Because the boundaries of ecosystems are not fixed, organisms and other ecosystem components can move freely between ecosystems. For example, the waste products from a terrestrial ecosystem may enter an aquatic ecosystem, changing its environmental characteristics. In addition, any ecosystem process that alters the global environment affects all the other ecosystems. For example, greenhouse gases that deplete the ozone layer and promote global warming affect the global climate, altering the characteristics of ecosystems worldwide.

Sample Test Questions and Rationale

112. **In the nitrogen cycle, decomposers are responsible for which process?**
     (Rigorous)

    A. Nitrogen fixing
    B. Nitrification
    C. Ammonification
    D. Assimilation

Answer: C. Ammonification

Nitrogen fixing and nitrification are primarily handled by bacteria (although inorganic processes augment this somewhat). Assimilation is a process performed by plants. Decomposers ammonify thus preparing nitrogen compounds for bacteria.

113. **Which biome is the most prevalent on Earth?**
     (Average Rigor)

    A. marine
    B. desert
    C. savanna
    D. tundra

Answer: A. Marine

The marine biome covers 75% of the Earth. This biome is organized by the depth of water.

TEACHER CERTIFICATION STUDY GUIDE

114. Which term is not associated with the water cycle?
 (Easy)

   A. precipitation
   B. transpiration
   C. fixation
   D. evaporation

Answer: C. fixation

Water is recycled through the processes of evaporation and precipitation. Transpiration is the evaporation of water from leaves. Fixation is not associated with the water cycle.

115. Which of the following terms does not describe a way that the human race negatively impacts the biosphere?
 (Rigorous)

   A. biological magnification
   B. pollution
   C. carrying capacity
   D. simplification of the food web

Answer: C. carrying capacity

Most people recognize the harmful effects of pollution, especially global warming. Pollution, and the regular use of pesticides and herbicides introduces toxins into the food web, biological magnification refers to the increasing concentration of these toxins as you move up the food web. Simplification of the food web has to do with small variety farming crops replacing large habitats, and thus shrinking or destroying some ecosystems. Carrying capacity on the other hand is simply a term for the amount of life a certain habitat can sustain, it is term independent of human action, so the answer is (C).

116. Which biogeochemical cycle plays the smallest part in photosynthesis or cellular respiration?
    (Rigorous)

    A. Hydrogen cycle
    B. Phosphorous cycle
    C. Sulfur cycle
    D. Nitrogen cycle

Answer: C. Sulfur cycle

Although sulfur is used in a small part of the proteins used in photosynthesis and respiration, only in some photosynthesizing bacteria (not cyanobacteria) does sulfur play a significant role. Hydrogen is readily passed through almost every stage of respiration and photosynthesis. Many of the energy storing molecules (including ATP) use phosphorous as a principle component. Nitrogen, in addition to being a significant component of proteins that assist in these processes, is also a significant component of NADP.

TEACHER CERTIFICATION STUDY GUIDE

DOMAIN 6.0 SCIENCE, TECHNOLOGY, AND SOCIETY

Competency 6.1 Impact of science and technology on the environment and human affairs

Many microorganisms are used to detoxify chemicals and recycle waste. Sewage treatment plants use microbes to degrade organic compounds. Some compounds, like chlorinated hydrocarbons, cannot be easily degraded. Scientists are currently researching ways to genetically modify microbes, thereby enabling them to degrade harmful compounds.

Genetic engineering has benefited agriculture also. For example, many dairy cows are given bovine growth hormone to increase milk production. Commercially grown plants are often genetically modified for optimal growth. Strains of wheat, cotton, and soybeans have been developed to resist herbicides that are used to control weeds. Crop plants are also being engineered to resist infections and pests. Scientists can genetically modify crops to contain viral genes that do not affect the plant and will "vaccinate" the plant from viral attack. Crop plants are now being modified to resist insect attacks as well allowing farmers to reduce the amount of pesticide used on plants.

Sample Test Question and Rationale

117. Genetic engineering is beneficial to agriculture in many ways. Which of the following is not an advantage of genetic engineering to agriculture? *(Average Rigor)*

    A. Use of bovine growth hormone to increase milk production.
    B. Development of crops resistant to herbicides.
    C. Development of microorganisms to breakdown toxic substances into harmless compounds.
    D. Genetic vaccination of plants against viral attack.

Answer: C. The development of microorganisms to breakdown toxic substances into harmless compounds.

All of the answers are actual results of genetic engineering, however only answer (C) has not been used for agricultural purposes. These microorganisms have however been used at toxic waster sites and oil spills.

TEACHER CERTIFICATION STUDY GUIDE

Competency 6.2 Hazards induced by humans and/or nature

An important topic in science is the effect of natural disasters and events on society and the effect human activity has on inducing such events. Naturally occurring geological, weather, and environmental events can greatly affect the lives of humans. In addition, the activities of humans can induce environmental events that would not normally occur.

Nature-induced hazards include floods, landslides, avalanches, volcanic eruptions, wildfires, earthquakes, hurricanes, tornadoes, droughts, and disease. Such events often occur naturally, because of changing weather patterns or geological conditions. Property damage, resource depletion, and the loss of human life are possible outcomes of natural hazards. Thus, natural hazards are often extremely costly on both an economic and personal level.

While many nature-induced hazards occur naturally, human activity can often stimulate such events. For example, destructive land use practices such as mining can induce landslides or avalanches if not properly planned and monitored. In addition, human activities can cause other hazards including global warming and waste contamination. Global warming is an increase in the Earth's average temperature resulting, at least in part, from the burning of fuels by humans.

Global warming is hazardous because it disrupts the Earth's environmental balance and can negatively affect weather patterns. The ecological and weather pattern changes induced by global warming promote natural disasters. Finally, improper hazardous waste disposal by humans can contaminate the environment. One important effect of hazardous waste contamination is the stimulation of disease in human populations. Thus, hazardous waste contamination negatively affects both the environment and the people that live in it.

Sample Test Questions and Rationale

118. **The biggest problem that humans have created in our environment is:** *(Average Rigor)*

    A. Nutrient depletion and excess.
    B. Global warming.
    C. Population overgrowth.
    D. Ozone depletion.

Answer: A. Nutrient depletion and excess.

Human activity affects parts of nutrient cycles by removing nutrients from one part of the biosphere and adding them to another. This results in nutrient depletion in one area and nutrient excess in another. This affects water systems, crops, wildlife, and humans. The answer is (A).

119. The demand for genetically enhanced crops has increased in recent years. Which of the following is not a reason for this increased demand?
(Easy)

   A. Fuel sources
   B. Increased growth
   C. Insect resistance
   D. Better-looking produce

Answer: D. Better-looking produce

Genetically enhanced crops are being developed for utilization as fuel sources, as well as for an increased production yield. Insect resistance eliminates the need for pesticides. While there may be some farmers crossing crops to make prettier watermelons, this is not a primary reason for the increased demand. The answer is (D).

Competency 6.3 Issues and applications (production, storage, use, management, and disposal of consumer products and energy, and management of natural resources)

An important application of science and technology is the production, storage, use, management, and disposal of consumer products and energy. Scientists from many disciplines work to produce a vast array of consumer products. Energy production and management is another area in which science plays a key role.

The production of a large number of popular consumer products requires scientific knowledge and technology. Genetically modified foods, pharmaceuticals, plastics, nylon, cosmetics, household cleaning products, and color additives are a few examples of science-based consumer goods. In addition to consumer product production, science helps determine the proper use and storage of consumer goods. Safe use and storage is a key component of successful production. For example, perishable products like food must be stored and used in a safe and sanitary way. Science helps establish limits and guidelines for the storage and use of perishable food products. Management and disposal of consumer products is also an important concern. Science helps establish limits for the safe use of potentially hazardous consumer products. For example, household cleaning products are potentially hazardous if used improperly. Scientific testing determines the proper uses and potential hazards of such products. Finally, disposal of waste from consumer product production and use is of great concern. Proper disposal of hazardous waste and recycling of durable materials is important for the health and safety of human populations and the long-term sustainability of the Earth's resources and environment.

Energy production and management is an increasingly important topic in scientific research because of the increasing scarcity of energy sources such as petroleum. With traditional sources of energy becoming more scarce and costly, a major goal of scientific energy research is the creation of alternative, efficient means of energy production. Examples of potential sources of alternative energy include wind, water, solar, nuclear, geothermal, and biomass. An important concern in the production and use of energy, from both traditional and alternative sources, is the effect on the environment and the safe disposal of waste products.

Scientific research helps determine the best method for energy production, use, and waste product disposal; balancing the need for energy with associated environmental and health concerns.

Sample Test Questions and Rationale

120. **Stewardship is the responsible management of resources. We must regulate our actions to do which of the following about environmental degradation?**
(Average Rigor)

    A. Prevent it.
    B. Reduce it.
    C. Mitigate it.
    D. All of the above.

Answer: D. All of the above.

Stewardship requires the regulation of human activity to prevent, reduce, and mitigate environmental degradation. An important aspect of stewardship is the preservation of resources and ecosystems for future human generations. Therefore, the answer is (D).

121. The three main concerns in nonrenewable resource management are conservation, environmental mitigation, and _____.
(Rigorous)

   A. preservation
   B. extraction
   C. allocation
   D. sustainability

Answer: C. allocation

The main concerns in nonrenewable resource management are conservation, allocation, and environmental mitigation. Policy makers, corporations, and governments must determine how to use and distribute scarce resources. Decision makers balance the immediate demand for resources with the need for resources in the future. This determination is often a cause of conflict and disagreement. Finally, scientists attempt to minimize and mitigate the environmental damage caused by resource extraction. The answer is (C).

Competency 6.4 Social, political, ethical, and economic issues in biology

Genetic engineering has drastically advanced biotechnology. Accompanying these advancements are several safety and ethical concerns. Many safety concerns have been abated by strict governmental regulations. The Food and Drug Administration (FDA), United States Department of Agriculture (USDA), Environmental Protection Agency (EPA), and National Institutes of Health (NIH) are just a few of the governmental agencies that regulate pharmaceutical, food, and environmental technology advancements.

Several ethical questions arise when discussing biotechnology. Should embryonic stem cell research be allowed? Is animal testing humane? These are just a couple of ethical questions that many people have. There are strong arguments for both sides and some governmental regulations are in place to monitor these issues.

Concepts that reflect a person's ethical viewpoints may be used for political purposes, either to further one's agenda, or to hurt an opponent. Recent political issues with ethical and scientific ties include abortion, stem cell research, and cloning. There are at least two sides to each issue and as such they can easily become partisan. Partisan delineation of ethical topics can greatly affect elections. Local, state, national, and global governments and organizations must increasingly consider policy issues related to science and technology. For example, local and state governments must analyze the impact of proposed development and growth on the environment. Governments and communities must balance the demands of an expanding human population with the local ecology to ensure sustainable growth.

TEACHER CERTIFICATION STUDY GUIDE

Another area affected by biology is economic health. Scientific and technological breakthroughs greatly influence other fields of study and the job market. All academic disciplines utilize computer and information technology to simplify research and information sharing. In addition, advances in science and technology influence the types of available jobs and the desired work skills of potential employees. For example, machines and computers continue to replace unskilled laborers and computer and technological literacy are now requirements for many jobs and careers. Finally, science and technology continue to change the very nature of careers. Because of science and technology's great influence on all areas of the economy, careers are far less stable than in past eras. Workers can thus expect to change jobs and companies much more frequently.

Sample Test Questions and Rationale

122. Which of the following limit the development of technological design ideas and solutions?

    I. monetary cost
    II. time
    III. laws of nature
    IV. governmental regulation
    (Average Rigor)

    A. I and II
    B. I, II, and IV
    C. II and III
    D. I, II, and III

Answer: D. I, II, and III

Technology cannot work against the laws of nature. Technological design solutions must work within the framework of the natural world. Monetary cost and time constraints also limit the development of new technologies. Governmental regulation, while present in many sciences, cannot regulate the formation of new ideas or design solutions. The answer, then, is D: I, II, and III.

TEACHER CERTIFICATION STUDY GUIDE

123. Which of the following is the least ethical choice for a school laboratory activity?
(Rigorous)

   A. Dissection of a donated cadaver.
   B. Dissection of a preserved fetal pig.
   C. Measuring the skeletal remains of birds.
   D. Pithing a frog to watch the circulatory system.

Answer: D. Pithing a frog to watch the circulatory system.

Scientific and societal ethics make choosing experiments in today's science classroom difficult. It is possible to ethically perform choices (A), (B), or (C), if due care is taken. (Note that students will need significant assistance and maturity to perform these experiments, and that due care also means attending to all legal paperwork that might be necessary.) However, modern practice precludes pithing animals (causing partial brain death while allowing some systems to function), as inhumane. Therefore, the answer to this question is (D).

Competency 6.5   Societal issues with health and medical advances

Society faces many issues associated with medical advances. Many medical advances require us to examine our beliefs. These ethical issues were addressed earlier. Another issue is the length of human life. As we develop cures for illnesses, and learn how to better care for ourselves and the environment, we are increasing our life expectancies. In the past the average person would live to be sixty and a family containing 13 children might only have two children reach adulthood. Many died from polio, measles, and mumps. Children today are routinely vaccinated for all three of these diseases. In the past, people were also more susceptible to death by infection, sometimes from a cut or abscessed tooth. Today, we have discovered more antibiotics to prevent prolonged biotic growth and subsequent illness.

These are common examples, but we have also found ways to cure less common, lethal diseases, and perform amazing surgical techniques, such as organ transplants. While many used to die from cancer, remission rates are now improving. Screening is essential and expensive, but not nearly as expensive as chemotherapy, radiation, and the drugs available for these diseases.

While it is wonderful to live a longer, healthier life, it has created challenges. Procedures that enable us to live longer are often costly. Many families do not have health insurance and are unable to pay for services, relying on state funded programs or foregoing services altogether. Of those families that do have health insurance, the cost of health insurance is continually increasing. A result is that some employers are no longer offering insurance as part of the employee benefit package. As we grow older and encounter more ailments, we require more medication to remain healthy. Many sources estimate the cost for a pharmaceutical company to bring a single drug to market at $350-$500 million. In order to recoup costs and remain profitable, companies must set the price of the drug accordingly, creating a loop of restricted access and unaffordable healthcare.

Sample Essays

1. Using your accumulated knowledge, discuss the components of biogeochemical cycles.

Best Response:

Essential elements are recycled through an ecosystem. At times, the element needs to be made available in a useable form. Cycles are dependent on plants, algae, and bacteria to fix nutrients for use by animals. The four main cycles are: water, carbon, nitrogen, and phosphorous.

Two percent of all the water is fixed in ice or the bodies of organisms, rendering it unavailable. Available water includes surface water (lakes, ocean, and rivers) and ground water (aquifers, wells). The majority (96%) of all available water is from ground water. Water is recycled through the processes of evaporation and precipitation. The water present now is the water that has been here since our atmosphere was formed.

Ten percent of all available carbon in the air (in the form of carbon dioxide gas) is fixed by photosynthesis. Plants fix carbon in the form of glucose. Animals eat the plants and are able to obtain the carbon necessary to sustain themselves. When animals release carbon dioxide through respiration, the cycle begins again as plants recycle the carbon through photosynthesis.

Eighty percent of the atmosphere is in the form of nitrogen gas. Nitrogen must be fixed and taken out of the gaseous form to be incorporated into an organism. Only a few genera of bacteria have the correct enzymes to break the strong triple bond between nitrogen atoms. These special bacteria live within the roots of legumes (e.g., peas, beans, alfalfa) and add bacteria to the soil so it may be taken-up by the plant. Nitrogen is necessary in the building of amino acids and the nitrogenous bases of DNA.

Phosphorus exists as a mineral and is not found in the atmosphere. Fungi and plant roots have structures called mycorrhizae that are able to fix insoluble phosphates into useable phosphorus. Urine and decayed matter returns phosphorus to the earth where it can be fixed in the plant. Phosphorus is needed for the backbone of DNA and for the manufacture of ATP.

The four biogeochemical cycles are present concurrently. Water is continually recycled, and is utilized by organisms to sustain life. Carbon is also a necessary component for life. Both water and carbon can be found in the air and on the ground. Nitrogen and phosphorous are commonly found in the ground. Special organisms, called decomposers, help to make these elements available in the environment. Plants use the recycled materials for energy and, when they are consumed, the cycle begins again.

Good Response:

Essential elements are recycled through an ecosystem. Cycles are dependent on plants, algae, and bacteria to make nutrients available for use by animals. The four main cycles are: water, carbon, nitrogen, and phosphorous. Water is typically available as surface water (large bodies of water) or ground water. Water is recycled through the states of gas, liquid (rain), and solid (ice or snow). Carbon is necessary for life as it is the basis for organic matter. It is a byproduct of photosynthesis and is found in the air as carbon dioxide gas. Nitrogen is the largest component of the atmosphere. It is also necessary for the creation of amino acids and the nitrogenous bases of DNA. Phosphorous is another elemental cycle. Phosphorous is found in the soil and is made available by decomposition. It is then converted for use in the manufacture of DNA and ATP.

Basic Response:

Elements are recycled through an ecosystem. This occurs through cycles. These important cycles are called biogeochemical cycles. The water cycle consists of water moving from bodies of water into the air and back again as precipitation. The carbon cycle includes all organisms, as mammals breathe out carbon dioxide and are made of carbon molecules. Nitrogen is an amino acid building block and is found in soil. As things are broken down, phosphorous is added to the earth, enriching the soil.

2. Examine the components of a eukaryotic cell.

Best Response:

The cell is the basic unit of all living things. Eukaryotic cells are found in protists, fungi, plants, and animals. Eukaryotic cells are organized. They contain many organelles, which are membrane bound areas for specific functions. Their cytoplasm contains a cytoskeleton that provides a protein framework for the cell. The cytoplasm also supports the organelles and contains the ions and molecules necessary for cell function. The cytoplasm is contained by the plasma membrane. The plasma membrane allows molecules to pass in and out of the cell. The membrane can bud inward to engulf outside material in a process called endocytosis. Exocytosis is a secretory mechanism, the reverse of endocytosis.

Eukaryotes have a nucleus. The nucleus is the brain of the cell that contains all of the cell's genetic information. The genetic information is contained on chromosomes that consist of chromatin, which are complexes of DNA and proteins. The chromosomes are tightly coiled to conserve space while providing a large surface area. The nucleus is the site of transcription of the DNA into RNA. The nucleolus is where ribosomes are made. There is at least one of these dark-staining bodies inside the nucleus of most eukaryotes.

The nuclear envelope is two membranes separated by a narrow space. The envelope contains many pores that let RNA out of the nucleus.

Ribosomes are the site for protein synthesis. They may be free floating in the cytoplasm or attached to the endoplasmic reticulum. There may be up to a half a million ribosomes in a cell, depending on how much protein the cell makes.

The endoplasmic reticulum (ER) is folded and provides a large surface area. It is the "roadway" of the cell and allows for transport of materials through and out of the cell. There are two types of ER: smooth and rough. Smooth endoplasmic reticulum contains no ribosomes on their surface. This is the site of lipid synthesis. Rough endoplasmic reticulum has ribosomes on its surfaces. They aid in the synthesis of proteins that are membrane bound or destined for secretion.

Many of the products made in the ER proceed to the Golgi apparatus. The Golgi apparatus functions to sort, modify, and package molecules that are made in the other parts of the cell. These molecules are either sent out of the cell or to other organelles within the cell. The Golgi apparatus is a stacked structure to increase the surface area.

Lysosomes are found mainly in animal cells. These contain digestive enzymes that break down food, unnecessary substances, viruses, damaged cell components, and eventually the cell itself. It is believed that lysosomes play a role in the aging process.

Mitochondria are large organelles that are the site of cellular respiration, where ATP is made to supply energy to the cell. Muscle cells have many mitochondria because they use a great deal of energy. Mitochondria have their own DNA, RNA, and ribosomes and are capable of reproducing by binary fission if there is a great demand for additional energy. Mitochondria have two membranes: a smooth outer membrane and a folded inner membrane. The folds inside the mitochondria are called cristae. They provide a large surface area for cellular respiration to occur.

Plastids are found only in photosynthetic organisms. They are similar to the mitochondria due to the double membrane structure. They also have their own DNA, RNA, and ribosomes and can reproduce if the need for the increased capture of sunlight becomes necessary. There are several types of plastids. Chloroplasts are the site of photosynthesis. The stroma is the chloroplast's inner membrane space. The stoma encloses sacs called thylakoids that contain the photosynthetic pigment chlorophyll. The chlorophyll traps sunlight inside the thylakoid to generate ATP, which is used in the stroma to produce carbohydrates and other products. The chromoplasts make and store yellow and orange pigments. They provide color to leaves, flowers, and fruits. The amyloplasts store starch and are used as a food reserve. They are abundant in roots like potatoes.

The Endosymbiotic Theory states that mitochondria and chloroplasts were once free living and possibly evolved from prokaryotic cells. At some point in our evolutionary history, they entered the eukaryotic cell and maintained a symbiotic relationship with the cell, with both the cell and organelle benefiting from the relationship. The fact that they both have their own DNA, RNA, ribosomes, and are capable of reproduction helps to confirm this theory.

Found in plant cells only, the cell wall is composed of cellulose and fibers. It is thick enough for support and protection, yet porous enough to allow water and dissolved substances to enter. Vacuoles are found mostly in plant cells. They hold stored food and pigments. Their large size allows them to fill with water in order to provide turgor pressure. Lack of turgor pressure causes a plant to wilt.

The cytoskeleton, found in both animal and plant cells, is composed of protein filaments attached to the plasma membrane and organelles. They provide a framework for the cell and aid in cell movement. They constantly change shape and move about. Three types of fibers make up the cytoskeleton:

1. Microtubules – the largest of the three, they make up cilia and flagella for locomotion. Some examples are sperm cells, cilia that line the fallopian tubes, and tracheal cilia. Centrioles are also composed of microtubules. They aid in cell division to form the spindle fibers that pull the cell apart into two new cells. Centrioles are not found in the cells of higher plants.

2. Intermediate filaments – intermediate in size, they are smaller than microtubules but larger than microfilaments. They help the cell to keep its shape.

3. Microfilaments – smallest of the three, they are made of actin and small amounts of myosin (like in muscle tissue). They function in cell movement like cytoplasmic streaming, endocytosis, and amoeboid movement. This structure pinches the two cells apart after cell division, forming two new cells.

Good Response:

The cell is the basic unit of all living things. Eukaryotic cells are found in protists, fungi, plants, and animals. Eukaryotic cells are organized. Their cytoplasm contains a cytoskeleton that provides a protein framework for the cell. The cytoplasm is contained by the plasma membrane. The plasma membrane allows molecules to pass in and out of the cell.

Eukaryotes have a nucleus. The nucleus is the brain of the cell that contains all of the cell's genetic information. The chromosomes house genetic information and are tightly coiled to conserve space while providing a large surface area. The nucleus is the site of transcription of the DNA into RNA. The nucleolus is where ribosomes are made.

Ribosomes are the site for protein synthesis. There may be up to a half a million ribosomes in a cell, depending on how much protein is made by the cell.

The endoplasmic reticulum (ER) is folded and provides a large surface area. It is the "roadway" of the cell and allows for transport of materials through and out of the cell. It may be smooth or rough.

Many of the products made in the ER proceed to the Golgi apparatus. The Golgi apparatus functions to sort, modify, and package molecules that are made in the other parts of the cell.

Mitochondria are large organelles that are the site of cellular respiration, where ATP is made to supply energy to the cell. Mitochondria have their own DNA, RNA, and ribosomes and are capable of reproducing by binary fission if there is a greater demand for additional energy.

Plastids are found only in photosynthetic organisms. They are similar to the mitochondria. They also have their own DNA, RNA, and ribosomes and can reproduce if the need for the increased capture of sunlight becomes necessary.

Found in plant cells only, the cell wall is composed of cellulose and fibers. It is thick enough for support and protection, yet porous enough to allow water and dissolved substances to enter.

The cytoskeleton, found in both animal and plant cells, is composed of protein filaments attached to the plasma membrane and organelles. They provide a framework for the cell and aid in cell movement. They constantly change shape and move about. Three types of fibers make up the cytoskeleton (in order of size, largest to smallest): microtubules, intermediate filaments, and microfilaments.

Basic Response:

The cell is the basic unit of all living things. Eukaryotic cells contain many organelles. Eukaryotes have a nucleus. The nucleus is the brain of the cell that contains all of the cell's genetic information. The nucleus is the site of DNA transcription. There is at least one nucleolus inside the nucleus of most eukaryotes. Ribosomes are the site for protein synthesis and can be found on the endoplasmic reticulum (ER). The Golgi apparatus functions to sort, modify, and package molecules that are made in the other parts of the cell. Mitochondria are large organelles that are the site of cellular respiration, where ATP is made to supply energy to the cell.

In plant cells, the cell wall is composed of cellulose and fibers. The cytoskeleton, found in both animal and plant cells, is composed of protein filaments. The three types of fibers differ in size and help the cell to keep its shape and aid in movement.

TEACHER CERTIFICATION STUDY GUIDE

3. Discuss the scientific process.

Best Response:

Science is a body of knowledge that is systematically derived from study, observations, and experimentation. Its goal is to identify and establish principles and theories that may be applied to solve problems. Pseudoscience, on the other hand, is a belief that is not warranted. There is no scientific methodology or application. Some of the more classic examples of pseudoscience include witchcraft, alien encounters, or any topic that is explained by hearsay.

Scientific experimentation must be repeatable. Experimentation leads to theories that can be disproved and are changeable. Science depends on communication, agreement, and disagreement among scientists. It is composed of theories, laws, and hypotheses.

A theory is the formation of principles or relationships that have been verified and accepted.

A law is an explanation of events that occur with uniformity under the same conditions (e.g., laws of nature, law of gravitation).

A hypothesis is an unproved theory or educated guess used to explain a phenomenon. A theory is a proven hypothesis.

Science is limited by the available technology. An example of this would be the relationship of the discovery of the cell to the invention of the microscope. As our technology improves, more hypotheses will become theories and possibly laws. Science is also limited by the data that we can collect. Data may be interpreted differently on different occasions. Science limitations cause explanations to be changeable as new technologies emerge.

The first step in scientific inquiry is posing a question. Next, a hypothesis is formed to provide a plausible explanation. An experiment is then proposed and performed to test this hypothesis. A comparison between the predicted and observed results is the next step. Conclusions are then formed and it is determined whether the hypothesis is correct or incorrect. If incorrect, the next step is to form a new hypothesis and repeat the process.

Good Response:

Science is derived from study, observations, and experimentation. The goal of science is to identify and establish principles and theories that may be applied to solve problems. Scientific theory and experimentation must be repeatable. It is also possible to disprove or change a theory. Science depends on communication, agreement, and disagreement among scientists. It is composed of theories, laws, and hypotheses.

A theory is a principle or relationship that has been verified and accepted through experiments. A law is an explanation of events that occur with uniformity under the same conditions. A hypothesis is an educated guess followed by research. A theory is a proven hypothesis.

Science is limited by the available technology. An example of this would be the relationship of the discovery of the cell to the invention of the microscope. The first step in scientific inquiry is posing a question. Next, a hypothesis is formed to provide a plausible explanation. An experiment is then proposed and performed to test this hypothesis. A comparison between the predicted and observed results is the next step. Conclusions are then formed and it is determined whether the hypothesis is correct or incorrect. If incorrect, the next step is to form a new hypothesis and repeat the process.

Basic Response:

Science is composed of theories, laws, and hypotheses. The first step in scientific inquiry is posing a question. Next, a hypothesis is formed to provide a plausible explanation. An experiment is then proposed and performed to test this hypothesis. A comparison between the predicted and observed results is the next step. Conclusions are then formed and it is determined whether the hypothesis is correct or incorrect. If incorrect, the next step is to form a new hypothesis and repeat the process. Science is always limited by the available technology.

Sample Test

Directions: Read each item and select the best response.

1. Identify the control in the following experiment. A student grew four plants under the following conditions and measured photosynthetic rate by measuring mass. He grew two plants in 50% light and two plants in 100% light.
   (Average Rigor, Skill 1.1.1)

   A. plants grown with no added nutrients.
   B. plants grown in the dark.
   C. plants grown in 100% light.
   D. plants grown in 50% light.

2. The concept that the rate of a given process is controlled by the most scarce factor in the process is known as? *(Average Rigor, Skill 1.1.5)*

   A. The Rate of Origination
   B. The Law of the Minimum
   C. The Law of Limitation
   D. The Law of Conservation

3. Which of the following discovered penicillin? *(Rigorous, Skill 1.1.9)*

   A. Pierre Curie
   B. Becquerel
   C. Louis Pasteur
   D. Alexander Fleming

4. The International System of Units (SI) measurement for temperature is on the _____ scale.
   (Rigorous, Skill 1.2.1)

   A. Celsius
   B. Farenheit
   C. Kelvin
   D. Rankine

5. In a data set, the value that occurs with the greatest frequency is referred to as the...
   (Average Rigor, Skill 1.2.3)

   A. mean.
   B. median.
   C. mode.
   D. range.

6. Three plants were grown and the following data recorded. Determine the mean growth. *(Easy, Skill 1.2.3)*

   Plant 1: 10 cm
   Plant 2: 20 cm
   Plant 3: 15 cm

   A. 5 cm
   B. 45 cm
   C. 12 cm
   D. 15 cm

7. In which of the following situations would a linear extrapolation of data be appropriate?
   (Rigorous, Skill 1.2.5)

   A. Computing the death rate of an emerging disease.
   B. Computing the number of plant species in a forest over time.
   C. Computing the rate of diffusion with a constant gradient.
   D. Computing the population at equilibrium.

8. Which of the following is not usually found on the MSDS for a laboratory chemical?
   (Rigorous, Skill 1.3.1)

   A. Melting Point
   B. Toxicity
   C. Storage Instructions
   D. Cost

9. Paper chromatography is most often associated with the separation of...
   (Average Rigor, Skill 1.3.2)

   A. nutritional elements.
   B. DNA.
   C. proteins.
   D. plant pigments.

10. Which item should always be used when using chemicals with noxious vapors?
    (Easy, Skill 1.3.3)

    A. Eye protection
    B. Face shield
    C. Fume hood
    D. Lab apron

11. Negatively charged particles that circle the nucleus of an atom are called...
    (Easy, Skill 2.1.1)

    A. neutrons.
    B. neutrinos.
    C. electrons.
    D. protons.

12. Which of the following are properties of water?

    I. High specific heat
    II. Strong ionic bonds
    III. Good solvent
    IV. High freezing point
    (Average Rigor, Skill 2.1.2)

    A. I, III, IV
    B. II and III
    C. I and II
    D. II, III, IV

13. Potassium chloride is an example of a(n):
    (Average Rigor, Skill 2.1.2)

    A. nonpolar covalent bond.
    B. polar covalent bond.
    C. ionic bond.
    D. hydrogen bond.

14. Sulfur oxides and nitrogen oxides in the environment react with water to cause:
    (Easy, Skill 2.1.3)

    A. ammonia.
    B. acidic precipitation.
    C. sulfuric acid.
    D. global warming.

15. What is necessary for diffusion to occur?
    (*Average Rigor, Skill 2.1.4*)

    E. Carrier proteins
    F. Energy
    G. A concentration gradient
    H. A membrane

16. The loss of an electron is _____ and the gain of an electron is _____.
    (*Rigorous, Skill 2.1.8*)

    A. oxidation, reduction
    B. reduction, oxidation
    C. glycolysis, photosynthesis
    D. photosynthesis, glycolysis

17. During the Krebs cycle, 8 carrier molecules are formed. What are they?
    (*Rigorous, Skill 2.1.8*)

    A. 3 NADH, 3 FADH, 2 ATP
    B. 6 NADH and 2 ATP
    C. 4 $FADH_2$ and 4 ATP
    D. 6 NADH and 2 $FADH_2$

18. The product of anaerobic respiration in animals is...
    (*Average Rigor, Skill 2.1.8*)

    E. Carbon dioxide.
    F. Lactic acid.
    G. Pyruvate.
    H. Ethyl alcohol.

19. ATP is known to bind to phosphofructokinase-1 (an enzyme involved in glycolysis). This results in a change in the shape of the enzyme that causes the rate of ATP production to fall. Which answer best describes this phenomenon?
    (*Rigorous, Skill 2.1.9*)

    A. Binding of a coenzyme
    B. An allosteric change in the enzyme
    C. Competitive inhibition
    D. Uncompetitive inhibition

20. The shape of a cell depends on its...
    (*Average Rigor, Skill 2.2.1*)

    A. function.
    B. structure.
    C. age.
    D. size.

21. Which type of cell would contain the most mitochondria?
    (*Average Rigor, Skill 2.2.1*)

    A. Muscle cell
    B. Nerve cell
    C. Epithelial cell
    D. Blood cell

22. Which of the follow is not true of both chloroplasts and mitochondria?
    (Easy, Skill 2.2.1)

    A. The inner membrane is the primary site for it's activity.
    B. Converts energy from one form to another.
    C. Uses an electron transport chain.
    D. Is an important part of the carbon cycle.

23. Which part of the cell is responsible for lipid synthesis?
    (Rigorous, Skill 2.2.1)

    A. Golgi Apparatus
    B. Rough Endoplasmic Reticulum
    C. Smooth Endoplasmic Reticulum
    D. Lysosome

24. According to the fluid-mosaic model of the cell membrane, membranes are composed of…
    (Rigorous, Skill 2.2.1)

    A. a phospholipid bilayer with proteins embedded in the layers.
    B. one layer of phospholipids with cholesterol embedded in the layer
    C. two layers of protein with lipids embedded in the layers.
    D. DNA and fluid proteins.

25. A type of molecule not found in the membrane of an animal cell is…
    (Rigorous, Skill 2.2.1)

    A. phospholipid.
    B. protein.
    C. cellulose.
    D. cholesterol.

26. Which of the following is not considered evidence of Endosymbiotic Theory?
    (Rigorous, Skill 2.2.1)

    A. The presence of genetic material in mitochondria and plastids.
    B. The presence of ribosomes within mitochondria and plastids.
    C. The presence of a double layered membrane in mitochondria and plastids.
    D. The ability of mitochondria and plastids to reproduce.

27. Identify this stage of mitosis.

    (Average Rigor, Skill 2.2.2)

    A. Anaphase
    B. Metaphase
    C. Telophase
    D. Prophase

28. Which statement regarding mitosis is correct?
    (Easy, Skill 2.2.2)

    A. Diploid cells produce haploid cells for sexual reproduction.
    B. Sperm and egg cells are produced.
    C. Diploid cells produce diploid cells for growth and repair.
    D. It allows for greater genetic diversity.

29. This stage of mitosis includes cytokinesis or division of the cytoplasm and its organelles.
    (Average Rigor, Skill 2.2.2)

    A. anaphase
    B. interphase
    C. prophase
    D. telophase

30. Replication of chromosomes occurs during which phase of the cell cycle?
    (Average Rigor, Skill 2.2.2)

    A. prophase
    B. interphase
    C. metaphase
    D. anaphase

31. Which process(es) result(s) in a haploid chromosome number?
    (Easy, Skill 2.2.2)

    A. Both meiosis and mitosis
    B. Mitosis
    C. Meiosis
    D. Replication and division

32. Crossing over, which increases genetic diversity, occurs during which stage(s)?
    (Rigorous, Skill 2.2.2)

    A. Telophase II in meiosis.
    B. Metaphase in mitosis.
    C. Interphase in both mitosis and meiosis.
    D. Prophase I in meiosis.

33. DNA synthesis results in a strand that is synthesized continuously. This is the...
    (Average Rigor, Skill 2.3.2)

    A. lagging strand.
    B. leading strand.
    C. template strand.
    D. complementary strand.

34. Segments of DNA can be transferred from one organism to another through the use of which of the following?
    (Average Rigor, Skill 2.3.2)

    A. Bacterial plasmids
    B. Viruses
    C. Chromosomes from frogs
    D. Plant DNA

35. Which of the following is not a form of posttranscriptional processing?
    (Rigorous, Skill 2.3.3)

    A. 5' capping
    B. intron splicing
    C. polypeptide splicing
    D. 3' polyadenylation

36. This carries amino acids to the ribosome in protein synthesis.
    (Average Rigor, Skill 2.3.3)

    A. messenger RNA
    B. ribosomal RNA
    C. transfer RNA
    D. DNA

37. A DNA molecule has the sequence ACTATG. What is the anticodon of this molecule?
    (Rigorous, Skill 2.3.3)

    A. UGAUAC
    B. ACUAUG
    C. TGATAC
    D. ACTATG

38. Viruses are made of...
    (Easy, Skill 2.3.6)

    A. a protein coat surrounding nucleic acid.
    B. DNA, RNA, and a cell wall.
    C. nucleic acid surrounding a protein coat.
    D. protein surrounded by DNA.

39. Cancer cells divide extensively and invade other tissues. This continuous cell division is due to...
    (Rigorous, Skill 2.3.7)

    A. density dependent inhibition.
    B. density independent inhibition.
    C. chromosome replication.
    D. growth factors.

40. Which of the following is not a useful application of genetic engineering?
    (Rigorous, Skill 2.3.8)

    A. The creation of safer viral vaccines.
    B. The creation of bacteria that produce hormones for medical use.
    C. The creation of bacteria to breakdown toxic waste.
    D. The creation of organisms that are successfully being used as sources for alternative fuels.

41. Which of the following is a way that cDNA cloning has not been used?
    (Rigorous, Skill 2.3.8)

    A. To provide evidence for taxonomic organization.
    B. To study the mutations that lead to diseases such as hemophilia.
    C. To determine the structure of a protein.
    D. To understand methods of gene regulation.

42. A genetic engineering advancement in the medical field is...
    (Easy, Skill 2.3.8)

    A. gene therapy.
    B. pesticides.
    C. degradation of harmful chemicals.
    D. antibiotics.

43. The Law of Segregation defined by Mendel states…
    (Average Rigor, Skill 3.1.1)

    A. when sex cells form, the two alleles that determine a trait will end up on different gametes.
    B. only one of two alleles is expressed in a heterozygous organism.
    C. the allele expressed is the dominant allele.
    D. alleles of one trait do not affect the inheritance of alleles on another chromosome.

44. A child with type O blood has a father with type A blood and a mother with type B blood. The genotypes of the parents, respectively, would be which of the following?
    (Average Rigor, Skill 3.1.1)

    A. AA and BO
    B. AO and BO
    C. AA and BB
    D. AO and OO

45. Amniocentesis is…
    (Average Rigor, Skill 3.1.4)

    A. a non-invasive technique for detecting genetic disorders.
    B. a bacterial infection.
    C. extraction of amniotic fluid.
    D. removal of fetal tissue.

46. A woman has Pearson Syndrome, a disease caused by a mutation in mitochondrial DNA. Which of the following individuals would you expect to see the disease in?
    (Rigorous, Skill 3.1.5)

    I   Her Daughter
    II Her Son
    III Her Daughter's son
    IV Her Son's daughter

    A. I, III
    B. I, II, III
    C. II, IV
    D. I, II, III, IV

47. Which is not a possible effect of polyploidy?
    (Rigorous, Skill 3.1.5)

    A. More robust members of an animal species.
    B. The creation of cross species offspring.
    C. The creation of a new species.
    D. Cells that produce higher levels of desired proteins.

48.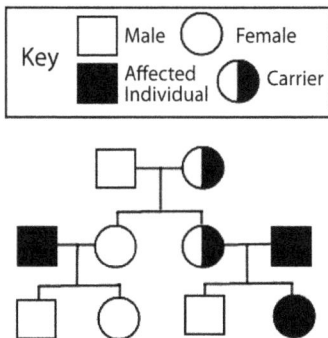

Based on the pedigree chart above, what term best describes the nature of the trait being mapped?
(Rigorous, Skill 3.1.5)

A. Autosomal recessive.
B. Sex-linked.
C. Incomplete dominance.
D. Co-dominance.

49. Which aspect of science does not support evolution?
(Average Rigor, Skill 3.2.1)

A. Comparative anatomy.
B. Organic chemistry.
C. Comparison of DNA among organisms.
D. Analogous structures.

50. Fossils of the dinosaur genus Saurolophus have been found in both Western North America and Mongolia. What is the most likely explanation for these findings?
(Rigorous, Skill 3.2.1)

A. Convergent evolution on the two continents lead to two species of dinosaurs with sufficient similarity to be placed in the same genus.
B. With the gaps in the fossil record, there are currently multiple competing theories to explain the presence of these fossils on two separate continents.
C. A distant ancestor of the these dinosaurs evolved before these land masses became separated.
D. Although Asia and North American were separate continents at the time, low sea levels made it possible for the dinosaurs to walk from one continent to the other.

51. **Which of the following best exemplifies the Theory of Inheritance of Acquired Traits?**
    (Average Rigor, Skill 3.2.2)

    A. Giraffes need to reach higher for leaves to eat, so their necks stretch. The giraffe babies are then born with longer necks. Eventually, there are more long-necked giraffes in the population.
    B. Giraffes with longer necks are able to reach more leaves, so they eat more and have more babies than other giraffes. Eventually, there are more long-necked giraffes in the population.
    C. Giraffes want to reach higher for leaves to eat, so they release enzymes into their bloodstream, which in turn causes fetal development of longer-necked giraffes. Eventually, there are more long-necked giraffes in the population.
    D. Giraffes with long necks are more attractive to other giraffes, so they get the best mating partners and produce more babies. Eventually, there are more long-necked giraffes in the population.

52. **Any change that affects the sequence of nucleotides in a gene is called a(n)...**
    (Easy, Skill 3.2.2)

    A. deletion.
    B. polyploid.
    C. mutation.
    D. duplication.

53. **The fossil record is often cited as providing evidence for the theory of punctuated equilibrium. Which of the following examples can only be explained by punctuated equilibrium and not gradualism?**
    (Rigorous, Skill 3.2.2)

    I   Coelacanth fish (once thought extinct) have remained relatively unchanged for millions of years.
    II The sudden apearance of a large number of different soft-bodied animals around 530 million years ago.
    III 10 million year old fossils and modern ginko plants are nearly identical.
    IV Fossils of Red Deer from the Island of Jersey show a six-fold decrease in body weight over the last 6000 years.

    A. I, III
    B. II, IV
    C. I, II, III, IV
    D. None of the above

54. **Evolution occurs in...**
    (Easy, Skill 3.2.3)

    A. individuals.
    B. populations.
    C. organ systems.
    D. cells.

55. Which of the following factors will affect the Hardy-Weinberg law of equilibrium, leading to evolutionary change?
    (Average Rigor, Skill 3.2.3)

    A. No mutations.
    B. Non-random mating.
    C. No immigration or emigration.
    D. Large population.

56. If a population is in Hardy-Weinbrium equilibrium and the frequency of the recessive allele is 0.3, what percentage of the population would be expected to be heterozygous?
    (Rigorous, Skill 3.2.3)

    A. 9%
    B. 49%
    C. 42%
    D. 21%

57. An animal choosing its mate because of attractive plumage or a strong mating call is an example of...
    (Average Rigor, Skill 3.2.3)

    A. sexual selection.
    B. natural selection.
    C. mechanical isolation.
    D. linkage.

58. Which of the following is an example of a phenotype that gives the organism an advantage in their home environment?
    (Rigorous, Skill 3.2.3)

    A. The color of Pepper Moths in England
    B. Thornless roses in a nature perserve
    C. Albinism in naked mole rats
    D. The large thorax of a Mediterranean fruit fly

59. Members of the same species:
    (Easy, Skill 3.2.4)

    A. look identical.
    B. never change.
    C. reproduce successfully within their group.
    D. live in the same geographic location.

60. Reproductive isolation results in...
    (Average Rigor, Skill 3.2.4)

    A. extinction.
    B. migration.
    C. follilization.
    D. speciation.

61. The biological species concept applies to
    (Average Rigor, Skill 3.2.4)

    A. asexual organisms.
    B. extinct organisms.
    C. sexual organisms.
    D. fossil organisms.

62. The first cells that evolved on earth were probably of which type?
    (Easy, Skill 3.2.6)

    A. Autotrophs.
    B. Eukaryotes.
    C. Heterotrophs.
    D. Prokaryotes.

63. All of the following gasses made up the primitive atmosphere except...
    (Average Rigor, Skill 3.2.6)

    A. ammonia
    B. methane
    C. oxygen
    D. hydrogen

64. Man's scientific name is Homo sapiens. Choose the proper classification beginning with kingdom and ending with order.
    (Average Rigor, Skill 4.1.1)

    A. Animalia, Vertebrata, Mammalia, Primate, Hominidae
    B. Animalia, Vertebrata, Chordata, Homo, sapiens
    C. Animalia, Chordata, Vertebrata, Mammalia, Primate
    D. Chordata, Vertebrata, Primate, Homo, sapiens

65. The two major ways to determine taxonomic classification are...
    (Average Rigor, Skill 4.1.1)

    A. evolution and phylogeny.
    B. reproductive success and evolution.
    C. phylogeny and morphology.
    D. size and color.

66. The scientific name Canis familiaris refers to the animal's...
    (Easy, Skill 4.1.1)

    A. kingdom and phylum names.
    B. genus and species names.
    C. class and species names.
    D. order and family names.

67. Thermoacidophiles are...
    (Average Rigor, Skill 4.1.2)

    A. prokaryotes.
    B. eukaryotes.
    C. protists.
    D. archaea.

68. Protists are classified into major groups according to...
    (Average Rigor, Skill 4.1.2)

    A. their method of obtaining nutrition.
    B. reproduction.
    C. metabolism.
    D. their form and function.

69. All of the following are examples of a member of Kingdom Fungi except...
    (Easy, Skill 4.1.2)

    A. mold.
    B. algae.
    C. mildew.
    D. mushrooms.

70. Which kingdom is comprised of organisms made of one cell with no nuclear membrane?
    (Easy, Skill 4.1.2)

    A. Monera
    B. Protista
    C. Fungi
    D. Algae

71. Within the Phylum Mollusca there are examples of both open and closed circulatory systems. Which of the following is a feature that is not common to both the open and closed cirulatory systems of molluscs?
    (Rigorous, Skill 4.1.2)

    A. Hemocoel
    B. Plasma
    C. Vessels
    D. Heart

72. Which of the following systems considers Archea (or Archeabacteria) as the most inclusive level of the taxonomic system?

    I   Three Domain System
    II Five Kingdom System
    III Six Kingdom System
    IV Eight Kingdom System
    (Rigorous, Skill 4.1.2)

    A. II, III
    B. I, IV
    C. I, III, IV
    D. I, II, III, IV

73. Laboratory researchers distinguish and classify fungi from plants because the cell walls of fungi contain _____.
    (Rigorous, Skill 4.1.2)

    A. chitin.
    B. lignin.
    C. lipopolysaccharides.
    D. cellulose.

74.

    Identify the correct characteristics of the plant pictured above.
    (Rigorous, Skill 4.2.1)

    A. seeded, non-vascular
    B. non-seeded, vascular
    C. non-seeded, non-vascular
    D. seeded, vascular

75. Which of the following is a characteristic of a monocot? *(Rigorous, Skill 4.2.1)*

   A. parallel veins in leaves
   B. flower petals occur in multiples of 4 or 5
   C. two seed leafs
   D. vascular tissue absent from the stem

76. Spores are the reproductive mode for which of the following group of plants? *(Average Rigor, Skill 4.2.1)*

   A. algae
   B. flowering plants
   C. conifers
   D. ferns

77.

   Using the following taxonomic key identify the tree that this branch came from?

   1 - Are the leaves PALMATELY COMPOUND (BLADES arranged like fingers on a hand)? – go to question 2
   1 - Are the leaves PINNATELY COMPOUND (BLADES arranged like the vanes of a feather)? – go to question 3

   2 - Are there usually 7 BLADES - Aesculus hippocastanum
   2 - Are there usually 5 BLADES - Aesculus glabra

   3 - Are there mostly 3-5 BLADES that are LOBED or coarsely toothed? - Acer negundo
   3 - Are there mostly 5-13 BLADES with smooth or toothed edges? - Fraxinus Americana
   (Rigorous, Skill 4.2.2)

   A. Aesculus hippocastanum
   B. Aesculus glabra
   C. Acer negundo
   D. Fraxinus Americana

78. Which of the following is not a factor that affects the rate of both photosynthesis and respiration in plants?
    (Average Rigor, Skill 4.2.3)

    A. the concentration of NADP and FAD
    B. the temperature
    C. the structure of plants
    D. the availability of different substrates

79. Oxygen is given off in the:
    (Easy, Skill 4.2.3)

    A. light reaction of photosynthesis.
    B. dark reaction of photosynthesis.
    C. Krebs cycle.
    D. reduction of NAD^+ to NADH.

80. The most ATP is generated through...
    (Rigorous, Skill 4.2.3)

    A. fermentation.
    B. glycolysis.
    C. chemiosmosis.
    D. the Krebs cycle.

81. Which of the following is not employed by a young cactus to survive in an arid environment?
    (Rigorous, Skill 4.2.3)

    A. Stem as the principle site of photosynthesis.
    B. A deep root system to reach additional sources of groundwater.
    C. CAM cycle photosynthesis.
    D. Spherical growth form.

82. Oxygen created in photosynthesis comes from the breakdown of
    (Average Rigor, Skill 4.2.3)

    A. carbon dioxide.
    B. water.
    C. glucose.
    D. carbon monoxide.

83. A plant cell is placed in salt water. The resulting movement of water out of the cell is called...
    (Average Rigor, Skill 4.2.3)

    A. facilitated diffusion.
    B. diffusion.
    C. transpiration.
    D. osmosis.

84. Double fertilization refers to which of the following?
    (Average Rigor, Skill 4.2.4)

    A. two sperm fertilizing one egg
    B. fertilization of a plant by gametes from two separate plants
    C. two sperm enter the plant embryo sac; one sperm fertilizes the egg, the other forms the endosperm
    D. the production of non-identical twins through fertilization of two separate eggs

85. The process in which pollen grains are released from the anthers is called:
    (Easy, Skill 4.2.4)

    A. pollination.
    B. fertilization.
    C. blooming.
    D. dispersal.

86. Which of the following is the correct order of the stages of plant development from egg to adult plant?
(Average Rigor, Skill 4.2.4)

   A. morphogenesis, growth, and cellular differentiation
   B. cell differentiation, growth, and morphogenesis
   C. growth, morphogenesis, and cellular differentiation
   D. growth, cellular differentiation, and morphogenesis

87. In angiosperms, the food for the developing plant is found in which of the following structures?
(Average Rigor, Skill 4.2.4)

   A. ovule
   B. endosperm
   C. male gametophyte
   D. cotyledon

88. In a plant cell, telophase is described as...
(Rigorous, Skill 4.2.4)

   A. the time of chromosome doubling.
   B. cell plate formation.
   C. the time when crossing over occurs.
   D. cleavage furrow formation.

89. Which of the following is a disadvantage of budding compared to sexual reproduction?
(Rigorous, Skill 4.2.4)

   A. limited number of offspring
   B. inefficient
   C. limited genetic diversity
   D. expensive to the parent organism

90. Which phylum accounts for 85% of all animal species?
(Easy, Skill 4.3.1)

   A. Nematoda
   B. Chordata
   C. Arthropoda
   D. Cnidaria

91. Fats are broken down by which substance?
(Average Rigor, Skill 4.3.2)

   A. bile produced in the gall bladder
   B. lipase produced in the gall bladder
   C. glucagons produced in the liver
   D. bile produced in the liver

92. A boy had the chicken pox as a baby. He will most likely not get this disease again because of...
(Average Rigor, Skill 4.3.2)

   A. passive immunity
   B. vaccination.
   C. antibiotics.
   D. active immunity.

93. Movement is possible by the action of muscles pulling on
 (Average Rigor, Skill 4.3.2)

    A. skin.
    B. bones.
    C. joints.
    D. ligaments.

94. Hormones are essential for the regulation of reproduction. What organ is responsible for the release of hormones for sexual maturity?
 (Average Rigor, Skill 4.3.2)

    A. pituitary gland
    B. hypothalamus
    C. pancreas
    D. thyroid gland

95 Which of the following compounds is not needed for skeletal muscle contraction to occur?
 (Rigorous, Skill 4.3.2)

    A. glucose
    B. sodium
    C. acetylcholine
    D. Adenosine 5'-triphosphate

96. Which of the following hormones is most involved in the process of osmoregulation?
 (Rigorous, Skill 4.3.2)

    A. Antidiuretic Hormone.
    B. Melatonin.
    C. Calcitonin.
    D. Gulcagon.

97. Capillaries come into contact with a large surface of both the kidneys and the lungs, especially in relation to the volume of these organs. Which of the following is not consistent with both organs and their contact with capillaries.
 (Rigorous, Skill 4.3.2)

    A. small specialized sections of each organ contact capillaries
    B. A large branching system of tubes within the organ
    C. A large source of blood that is quickly divided into capillaries
    D. A sack that contains a capillary network

98. Which of the following substances in unlikely to cause negative consequences if over-ingested?
 (Rigorous, Skill 4.3.2)

    A. essential fatty acids
    B. essential minerals
    C. essential water-insoluble vitamins
    D. essential water-soluble vitamins

99. If someone were experiencing unexplained changes in body temperature, hunger, and circadian rythyms, which of the following structures would most likely be the cause of these problems?
 (Rigorous, Skill 4.3.2)

    A. hypothalamus
    B. central nervous system
    C. pineal gland
    D. basal ganglia

100. Which is the correct sequence of embryonic development in a frog?
    (Average Rigor, Skill 4.3.3)

    A. cleavage – blastula – gastrula
    B. cleavage – gastrula – blastula
    C. blastula – cleavage – gastrula
    D. gastrula – blastula – cleavage

101. Fertilization in humans usually occurs in the:
    (Easy, Skill 4.3.3)

    A. cervix.
    B. ovary.
    C. fallopian tubes.
    D. vagina.

102. Which of the following has no relation to female sexual maturity?
    (Rigorous, Skill 4.3.3)

    A. thyroxine
    B. estrogen
    C. testosterone
    D. luteinizing hormone

103. In the growth of a population, initially the increase is exponential until carrying capacity is reached. This is represented by a(n):
    (Average Rigor, Skill 5.1)

    A. S curve.
    B. J curve.
    C. M curve.
    D. L curve.

104. All of the following are density dependent factors that affect a population except
    (Rigorous, Skill 5.1)

    A. disease.
    B. drought.
    C. predation.
    D. migration.

105. Which of the following is not an example of dynamic equilibrium?
    (Rigorous, Skill 5.1)

    A. a stable population
    B. a symbiotic pair of organisms
    C. osmoregulation
    D. maintaining head position while walking

106. All of the following are density independent factors that affect a population except...
    (Average Rigor, Skill 5.1)

    A. temperature.
    B. rainfall.
    C. predation.
    D. soil nutrients.

107. If the niches of two species overlap, what usually results?
    (Easy, Skill 5.2)

    A. a symbiotic relationship
    B. cooperation
    C. competition
    D. a new species

108. Primary succession occurs after...
    (Average Rigor, Skill 5.2)

    A. nutrient enrichment.
    B. a forest fire.
    C. exposure of a bare rock after the water table permanently recedes.
    D. a housing development is built.

109. A clownfish is protected by the sea anemone's tentacles. In turn, the anemone receives uneaten food from the clownfish. This is an example of...
    (Easy, Skill 5.2)

    A. mutualism.
    B. parasitism.
    C. commensalism.
    D. competition.

110. Which of the following are reasons to maintain biological diversity?

    I. Consumer product development.
    II. Stability of the environment.
    III. Habitability of our planet.
    IV. Cultural diversity.
    (Rigorous, Skill 5.2)

    A. I and III
    B. II and III
    C. I, II, and III
    D. I, II, III, and IV

111. Which of the following is not an abiotic factor?
    (Easy, Skill 5.2)

    A. temperature
    B. rainfall
    C. soil quality
    D. bacteria

112. In the nitrogen cycle, decomposers are responsible for which process?
    (Rigorous, Skill 5.3)

    A. Nitrogen fixing
    B. Nitrification
    C. Ammonification
    D. Assimilation

113. Which biome is the most prevalent on Earth?
    (Average Rigor, Skill 5.3)

    A. marine
    B. desert
    C. savanna
    D. tundra

114. Which term is not associated with the water cycle?
    (Easy, Skill 5.3)

    A. precipitation
    B. transpiration
    C. fixation
    D. evaporation

115. Which of the following terms does not describe a way that the human race negatively impacts the biosphere?
    (Rigorous, Skill 5.3)

    A. biological magnification
    B. pollution
    C. carrying capacity
    D. simplification of the food web

116. Which biogeochemical cycle plays the smallest part in photosynthesis or cellular respiration?
    (Rigorous, Skill 5.3)

    A. Hydrogen cycle
    B. Phosphorous cycle
    C. Sulfur cycle
    D. Nitrogen cycle

117. Genetic engineering is beneficial to agriculture in many ways. Which of the following is not an advantage of genetic engineering to agriculture?
    (Average Rigor, Skill 6.1)

    A. Use of bovine growth hormone to increase milk production.
    B. Development of crops resistant to herbicides.
    C. Development of microorganisms to breakdown toxic substances into harmless compounds.
    D. Genetic vaccination of plants against viral attack.

118. The biggest problem that humans have created in our environment is:
    (Average Rigor, Skill 6.2)

    A. Nutrient depletion and excess.
    B. Global warming.
    C. Population overgrowth.
    D. Ozone depletion.

119. The demand for genetically enhanced crops has increased in recent years. Which of the following is not a reason for this increased demand?
    (Easy, Skill 6.2)

    A. Fuel sources
    B. Increased growth
    C. Insect resistance
    D. Better-looking produce

120. Stewardship is the responsible management of resources. We must regulate our actions to do which of the following about environmental degradation?
    (Average Rigor, Skill 6.3)

    A. Prevent it.
    B. Reduce it.
    C. Mitigate it.
    D. All of the above.

121. The three main concerns in nonrenewable resource management are conservation, environmental mitigation, and _____.
    (Rigorous, Skill 6.3)

    A. preservation
    B. extraction
    C. allocation
    D. sustainability

122. Which of the following limit the development of technological design ideas and solutions?

   I. monetary cost
   II. time
   III. laws of nature
   IV. governmental regulation
   (Average Rigor, Skill 6.4)

   A. I and II
   B. I, II, and IV
   C. II and III
   D. I, II, and III

123. Which of the following is the least ethical choice for a school laboratory activity?
   (Rigorous, Skill 6.4)

   A. Dissection of a donated cadaver.
   B. Dissection of a preserved fetal pig.
   C. Measuring the skeletal remains of birds.
   D. Pithing a frog to watch the circulatory system.

Answer Key

1.	C	46.	B	91.	D
2.	B	47.	A	92.	D
3.	D	48.	B	93.	B
4.	C	49.	B	94.	B
5.	C	50.	D	95.	A
6.	D	51.	A	96.	A
7.	C	52.	C	97.	D
8.	D	53.	D	98.	D
9.	D	54.	B	99.	A
10.	C	55.	B	100.	A
11.	C	56.	C	101.	C
12.	A	57.	A	102.	A
13.	C	58.	A	103.	A
14.	B	59.	C	104.	B
15.	C	60.	D	105.	D
16.	A	61.	C	106.	C
17.	D	62.	D	107.	C
18.	B	63.	C	108.	C
19.	B	64.	C	109.	A
20.	A	65.	C	110.	D
21.	A	66.	B	111.	D
22.	A	67.	D	112.	C
23.	C	68.	D	113.	A
24.	A	69.	B	114.	C
25.	C	70.	A	115.	C
26.	C	71.	A	116.	C
27.	B	72.	C	117.	C
28.	C	73.	A	118.	A
29.	D	74.	B	119.	D
30.	B	75.	A	120.	D
31.	C	76.	D	121.	C
32.	D	77.	C	122.	D
33.	B	78.	C	123.	D
34.	A	79.	A		
35.	C	80.	C		
36.	C	81.	B		
37.	B	82.	B		
38.	A	83.	B		
39.	B	84.	C		
40.	D	85.	A		
41.	C	86.	C		
42.	A	87.	B		
43.	A	88.	B		
44.	B	89.	C		
45.	C	90.	C		

TEACHER CERTIFICATION STUDY GUIDE

Rigor Table

	Easy 20%	Average Rigor 40%	Rigorous 40%
Question #	6, 10, 11, 14, 22, 28, 31, 38, 42, 52, 54, 59, 62, 66, 69, 70, 79, 85, 90, 101, 107, 109, 111, 114, 119	1, 2, 5, 9, 12, 13, 15, 18, 20, 21, 27, 29, 30, 33, 34, 36, 43, 44, 45, 49, 51, 55, 57, 60, 61, 63, 64, 65, 67, 68, 76, 78, 82, 83, 84, 86, 87, 91, 92, 93, 94, 100, 103, 106, 108, 113, 117, 118, 120, 122	3, 4, 7, 8, 16, 17, 19, 23, 24, 25, 26, 32, 35, 37, 39, 40, 41, 46, 47, 48, 50, 53, 56, 58, 71, 72, 73, 74, 75, 77, 80, 81, 88, 89, 95, 96, 97, 98, 99, 102, 104, 105, 110, 112, 115, 116, 121, 123

TEACHER CERTIFICATION STUDY GUIDE

Sample Questions with Rationale

1. **Identify the control in the following experiment. A student grew four plants under the following conditions and measured photosynthetic rate by measuring mass. He grew two plants in 50% light and two plants in 100% light.**
   (Average Rigor, Skill 1.1.1)

   A. plants grown with no added nutrients.
   B. plants grown in the dark.
   C. plants grown in 100% light.
   D. plants grown in 50% light.

Answer: C. plants grown in 100% light.

The plants grown in 100% light are the control that the student will compare the growth of the plants 50% light.

2. **The concept that the rate of a given process is controlled by the most scarce factor in the process is known as?**
   (Average Rigor, Skill 1.1.5)

   A. The Rate of Origination
   B. The Law of the Minimum
   C. The Law of Limitation
   D. The Law of Conservation

Answer: B. The Law of the Minimum

A limiting factor is the component of a biological process that determines how quickly or slowly the process proceeds. Photosynthesis is the main biological process determining the rate of ecosystem productivity or the rate at which an ecosystem creates biomass. Thus, in evaluating the productivity of an ecosystem, potential limiting factors are light intensity, gas concentrations, and mineral availability. The Law of the Minimum states that the required factor which is most scarce in a given process controls the rate of the process.

3. Which of the following discovered penicillin?
   (Rigorous, Skill 1.1.9)

   A. Pierre Curie
   B. Becquerel
   C. Louis Pasteur
   D. Alexander Fleming

Answer: D. Alexander Fleming

Sir Alexander Fleming was a pharmacologist from Scotland. He isolated the antibiotic penicillin from a fungus in 1928.

4. The International System of Units (SI) measurement for temperature is on the _____ scale.
   (Rigorous, Skill 1.2.1)

   A. Celsius
   B. Farenheit
   C. Kelvin
   D. Rankine

Answer: C. Kelvin

Science uses the SI system because of its worldwide acceptance and ease of comparison. The SI scale for measuring temperature is the Kelvin Scale. Science, however, uses the Celsius scale for its ease of use. The answer is (C).

5. In a data set, the value that occurs with the greatest frequency is referred to as the...
   (Average Rigor, Skill 1.2.3)

   A. mean.
   B. median.
   C. mode.
   D. range.

Answer: C. Mode

Mean is the mathematical average of all the items. The median depends on whether the number of items is odd or even. If the number is odd, then the median is the value of the item in the middle. Mode is the value of the item that occurs the most often, if there are not many items. Bimodal is a situation where there are two items with equal frequency. Range is the difference between the maximum and minimum values.

6. **Three plants were grown and the following data recorded. Determine the mean growth.** (*Easy, Skill 1.2.3*)

   Plant 1: 10 cm
   Plant 2: 20 cm
   Plant 3: 15 cm

   A. 5 cm
   B. 45 cm
   C. 12 cm
   D. 15 cm

Answer: D. 15 cm

The mean growth is the average of the three growth heights.

$$\frac{10 + 20 + 15}{3} = 15 \text{ cm average height}$$

7. **In which of the following situations would a linear extrapolation of data be appropriate?**
   (*Rigorous, Skill 1.2.5*)

   A. Computing the death rate of an emerging disease.
   B. Computing the number of plant species in a forest over time.
   C. Computing the rate of diffusion with a constant gradient.
   D. Computing the population at equilibrium.

Answer: C: Computing the rate of diffusion with a constant gradient.

The individual data points on a linear graph cluster around a line of best fit. In other words, a relationship is linear if we can sketch a straight line that roughly fits the data points. Extrapolation is the process of estimating data points outside a known set of data points. When extrapolating data of a linear relationship, we extend the line of best fit beyond the known values. The extension of the line represents the estimated data points. Extrapolating data is only appropriate if we are relatively certain that the relationship is indeed linear. The answer is (C).

TEACHER CERTIFICATION STUDY GUIDE

8. **Which of the following is not usually found on the MSDS for a laboratory chemical?**
   (Rigorous, Skill 1.3.1)

   A. Melting Point
   B. Toxicity
   C. Storage Instructions
   D. Cost

Answer: D. Cost

MSDS, or Material Safety Data Sheets, are used to make sure that anyone can easily obtain information about a chemical, especially in the event of a spill or accident. This information typically includes physical data, toxicity, health effects, first aid, reactivity, storage, disposal, protective measures, and spill/leak procedures. Cost is not generally included as MSDS's. Costs are generated by the distributor, and seperate suppliers may have different costs. The answer, therefore, is (D).

9. **Paper chromatography is most often associated with the separation of...**
   (Average Rigor, Skill 1.3.2)

   A. nutritional elements.
   B. DNA.
   C. proteins.
   D. plant pigments.

Answer: D. plant pigments.

Paper chromatography uses the principles of capillarity to separate substances such as plant pigments. Molecules of a larger size will move more slowly up the paper, whereas smaller molecules will move more quickly producing lines of pigment.

BIOLOGY

10. Which item should always be used when using chemicals with noxious vapors?
  (Easy, Skill 1.3.3)

  A. Eye protection
  B. Face shield
  C. Fume hood
  D. Lab apron

Answer: C. Fume hood

Fume hoods are designed to protect the experimenter from chemical fumes. The three other choices do not prevent chemical fumes from entering the respiratory system.

11. Negatively charged particles that circle the nucleus of an atom are called...
  (Easy, Skill 2.1.1)

  A. neutrons.
  B. neutrinos.
  C. electrons.
  D. protons.

Answer: C. electrons.

Neutrons and protons make up the core of an atom. Neutrons have no charge and protons are positively charged. Electrons are the negatively charged particles around the nucleus.

12. Which of the following are properties of water?

  I. High specific heat
  II. Strong ionic bonds
  III. Good solvent
  IV. High freezing point
  (Average Rigor, Skill 2.1.2)

  A. I, III, IV
  B. II and III
  C. I and II
  D. II, III, IV

Answer: A. I, III, IV

All are properties of water except strong ionic bonds. Water is held together by polar covalent bonds between hydrogen and oxygen atoms.

13. **Potassium chloride is an example of a(n):**
    (Average Rigor, Skill 2.1.2)

    A. nonpolar covalent bond.
    B. polar covalent bond.
    C. ionic bond.
    D. hydrogen bond.

Answer: C. ionic bond.

Ionic bonds are formed when one electron is stripped away from its atom to join another atom. Ionic compounds are called salts and potassium chloride is a salt; therefore, potassium chloride is an example of an ionic bond.

14. **Sulfur oxides and nitrogen oxides in the environment react with water to cause:**
    (Easy, Skill 2.1.3)

    A. ammonia.
    B. acidic precipitation.
    C. sulfuric acid.
    D. global warming.

Answer: B. acidic precipitation

Acidic precipitation is rain, snow, or fog with a pH less than 5.6. It is caused by sulfur oxides and nitrogen oxides that react with water in the air to form acids that fall down to Earth as precipitation.

15. **What is necessary for diffusion to occur?**
    (Average Rigor, Skill 2.1.4)

    I. Carrier proteins
    J. Energy
    K. A concentration gradient
    L. A membrane

Answer: C. A concentration gradient

Diffusion is the ability of molecules to move from areas of high concentration to areas of low concentration (a concentration gradient).

TEACHER CERTIFICATION STUDY GUIDE

16. The loss of an electron is _____ and the gain of an electron is _____.
    (Rigorous, Skill 2.1.8)

    A. oxidation, reduction
    B. reduction, oxidation
    C. glycolysis, photosynthesis
    D. photosynthesis, glycolysis

Answer: A. oxidation, reduction

Oxidation-reduction reactions are also known as redox reactions. In respiration, energy is released by the transfer of electrons in redox reactions. The oxidation phase of this reaction involve the loss of an electron and the reduction phase involves the gain of an electron.

17. During the Krebs cycle, 8 carrier molecules are formed. What are they?
    (Rigorous, Skill 2.1.8)

    A. 3 NADH, 3 FADH, 2 ATP
    B. 6 NADH and 2 ATP
    C. 4 FADH$_2$ and 4 ATP
    D. 6 NADH and 2 FADH$_2$

Answer: D. 6 NADH and 2 FADH$_2$

For each molecule of CoA that enters the Kreb's cycle, you get 3 NADH and 1 FADH$_2$. There are 2 molecules of CoA so the total yield is 6 NADH and 2 FADH$_2$ during the Krebs cycle.

18. The product of anaerobic respiration in animals is…
    (Average Rigor, Skill 2.1.8)

    I. carbon dioxide.
    J. lactic acid.
    K. pyruvate.
    L. ethyl alcohol.

Answer: B. lactic acid.

In anaerobic lactic acid fermentation, pyruvate is reduced by NADH to form lactic acid. This is the anaerobic process in animals. Alcoholic fermentation is an anaerobic process in yeast and some bacteria yielding ethyl alcohol. Carbon dioxide and pyruvate are products of aerobic respiration.

BIOLOGY

19. ATP is known to bind to phosphofructokinase-1 (an enzyme involved in glycolysis). This results in a change in the shape of the enzyme that causes the rate of ATP production to fall. Which answer best describes this phenomenon?
(Rigorous, Skill 2.1.9)

- A. Binding of a coenzyme
- B. An allosteric change in the enzyme
- C. Competitive inhibition
- D. Uncompetitive inhibition

Answer: B. An allosteric change in the enzyme

The binding of ATP to phosphofructokinase-1 causes an allosteric change (a change in shape) of the enzyme. The binding of ATP can be considered non-competitive inhibition.

20. The shape of a cell depends on its...
(Average Rigor, Skill 2.2.1)

- A. function.
- B. structure.
- C. age.
- D. size.

Answer: A. function.

In most living organisms, cellular structure is based on function.

21. Which type of cell would contain the most mitochondria?
(Average Rigor, Skill 2.2.1)

- A. Muscle cell
- B. Nerve cell
- C. Epithelial cell
- D. Blood cell

Answer: A. Muscle cell

Mitochondria are the site of cellular respiration where ATP is produced. Muscle cells have the most mitochondria because they use a great deal of energy.

22. **Which of the follow is not true of both chloroplasts and mitochondria?**
    (Easy, Skill 2.2.1)

    A. The inner membrane is the primary site for it's activity.
    B. Converts energy from one form to another.
    C. Uses an electron transport chain.
    D. Is an important part of the carbon cycle.

Answer: A. The inner membrane is the primary site for it's activity.

In mitochondria the electron transport chain is present in the inner membrane, however in chloroplasts it is present in the thylakoid membranes.

23. **Which part of the cell is responsible for lipid synthesis?**
    (Rigorous, Skill 2.2.1)

    A. Golgi Apparatus
    B. Rough Endoplasmic Reticulum
    C. Smooth Endoplasmic Reticulum
    D. Lysosome

Answer: C. Smooth Endoplasmic Reticulum

The rough endoplasmic reticulum and the golgi apparatus are both involved in the production of proteins (synthesis and packaging, respectively). Lysosomes contain digestive enzymes. Only the smooth endoplasmic reticulum is directly responsible for lipid production.

24. **According to the fluid-mosaic model of the cell membrane, membranes are composed of...**
    (Rigorous, Skill 2.2.1)

    A. a phospholipid bilayer with proteins embedded in the layers.
    B. one layer of phospholipids with cholesterol embedded in the layer
    C. two layers of protein with lipids embedded in the layers.
    D. DNA and fluid proteins.

Answer: A. phospholipid bilayers with proteins embedded in the layers.

Cell membranes are composed of a phospholipid bilayer in which hydrophilic heads face outward and hydrophobic tails are sandwiched between the hydrophilic layers. The membrane contains proteins embedded in the layer (integral proteins) and proteins on the surface (peripheral proteins).

25. **A type of molecule not found in the membrane of an animal cell is...**
    (Rigorous, Skill 2.2.1)

    A. phospholipid.
    B. protein.
    C. cellulose.
    D. cholesterol.

Answer: C. cellulose.

Phospholipids, protein, and cholesterol are all found in animal cells. Cellulose, however, is only found in plant cells.

26. **Which of the following is not considered evidence of Endosymbiotic Theory?**
    (Rigorous, Skill 2.2.1)

    A. The presence of genetic material in mitochondria and plastids.
    B. The presence of ribosomes within mitochondria and plastids.
    C. The presence of a double layered membrane in mitochondria and plastids.
    D. The ability of mitochondria and plastids to reproduce.

Answer: C. The presence of a double layered membrane in mitochondria and plastids.

A double layered membrane is not unique to mitochondria and plastids, the nucleus is also double layered. All three other characteristics are not present in any other organelle, and are evidence that mitochondria and plastids may once have been separate organisms.

27. Identify this stage of mitosis.

(Average Rigor, Skill 2.2.2)

    A. Anaphase
    B. Metaphase
    C. Telophase
    D. Prophase

Answer: B. Metaphase

During metaphase, the centromeres are at opposite ends of the cell. During this phase the chromosomes are aligned with one another in the middle of the cell.

28. Which statement regarding mitosis is correct?
(Easy, Skill 2.2.2)

    A. Diploid cells produce haploid cells for sexual reproduction.
    B. Sperm and egg cells are produced.
    C. Diploid cells produce diploid cells for growth and repair.
    D. It allows for greater genetic diversity.

Answer: C. Diploid cells produce diploid cells for growth and repair.

The purpose of mitotic cell division is to provide growth and repair in body (somatic) cells. The cells begin as diploid and produce diploid cells.

29. This stage of mitosis includes cytokinesis or division of the cytoplasm and its organelles.
(Average Rigor, Skill 2.2.2)

    A. Anaphase
    B. Interphase
    C. Prophase
    D. Telophase

Answer: D. Telophase

The last stage of the mitosis is telophase. Here, the two nuclei form with a full set of DNA each. The cell is pinched in half into two cells and cytokinesis, or the division of the cytoplasm and organelles, occurs.

30. **Replication of chromosomes occurs during which phase of the cell cycle?**
    (Average Rigor, Skill 2.2.2)

    A. Prophase
    B. Interphase
    C. Metaphase
    D. Anaphase

Answer: B. Interphase

Interphase is the stage where the cell grows and copies the chromosomes in preparation for the mitotic phase.

31. **Which process(es) result(s) in a haploid chromosome number?**
    (Easy, Skill 2.2.2)

    A. Both meiosis and mitosis
    B. Mitosis
    C. Meiosis
    D. Replication and division

Answer: C. Meiosis

In meiosis, there are two consecutive cell divisions resulting in the reduction of chromosome number by half (diploid to haploid).

32. **Crossing over, which increases genetic diversity, occurs during which stage(s)?**
    (Rigorous, Skill 2.2.2)

    A. Telophase II in meiosis.
    B. Metaphase in mitosis.
    C. Interphase in both mitosis and meiosis.
    D. Prophase I in meiosis.

Answer: D. Prophase I in meiosis.

During prophase I of meiosis, the replicated chromosomes condense and pair with their homologues in a process called synapsis. Crossing over, the exchange of genetic material between homologues, occurs during prophase I.

33. **DNA synthesis results in a strand that is synthesized continuously. This is the...**
    (Average Rigor, Skill 2.3.2)

    A. lagging strand.
    B. leading strand.
    C. template strand.
    D. complementary strand.

Answer: B. leading strand.

As DNA synthesis proceeds along the replication fork, one strand is replicated continuously (the leading strand) and the other strand is replicated discontinuously (the lagging strand).

34. **Segments of DNA can be transferred from one organism to another through the use of which of the following?**
    (Average Rigor, Skill 2.3.2)

    A. Bacterial plasmids
    B. Viruses
    C. Chromosomes from frogs
    D. Plant DNA

Answer: A. Bacterial plasmids

Plasmids can transfer themselves (and therefore their genetic information) through a process called conjugation. This requires cell-to-cell contact.

35. **Which of the following is not a form of posttranscriptional processing?**
    (Rigorous, Skill 2.3.3)

    A. 5' capping
    B. intron splicing
    C. polypeptide splicing
    D. 3' polyadenylation

Answer: C. polypeptide splicing

The removal of segments of polypeptides is a posttranslational process. The other three are methods of posttranscriptional processing.

36. **This carries amino acids to the ribosome in protein synthesis.**
    (Average Rigor, Skill 2.3.3)

    A. messenger RNA
    B. ribosomal RNA
    C. transfer RNA
    D. DNA

Answer: C. transfer RNA

The tRNA molecule is specific for a particular amino acid. The tRNA has an anticodon sequence that is complementary to the codon. This specifies where the tRNA places the amino acid in protein synthesis.

37. **A DNA molecule has the sequence ACTATG. What is the anticodon of this molecule?**
    (Rigorous, Skill 2.3.3)

    A. UGAUAC
    B. ACUAUG
    C. TGATAC
    D. ACTATG

Answer: B. ACUAUG

The DNA is first transcribed into mRNA. Here, the DNA has the sequence ACTATG; therefore, the complementary mRNA sequence is UGAUAC (remember, in RNA, T is U). This mRNA sequence is the codon. The anticodon is the complement to the codon. The anticodon sequence will be ACUAUG (remember, the anticodon is tRNA, so U is present instead of T).

38. **Viruses are made of...**
    (Easy, Skill 2.3.6)

    A. a protein coat surrounding nucleic acid.
    B. DNA, RNA, and a cell wall.
    C. nucleic acid surrounding a protein coat.
    D. protein surrounded by DNA.

Answer: A. a protein coat surrounding nucleic acid.

Viruses are composed of a protein coat and nucleic acid; either RNA or DNA.

39. Cancer cells divide extensively and invade other tissues. This continuous cell division is due to…
    (Rigorous, Skill 2.3.7)

    A. density dependent inhibition.
    B. density independent inhibition.
    C. chromosome replication.
    D. growth factors.

Answer: B. density independent inhibition.

Density dependent inhibition is when the cells crowd one another and consume all the nutrients, thereby halting cell division. Cancer cells, however, are density independent; meaning they can divide continuously as long as nutrients are present.

40. Which of the following is not a useful application of genetic engineering?
    (Rigorous, Skill 2.3.8)

    A. The creation of safer viral vaccines.
    B. The creation of bacteria that produce hormones for medical use.
    C. The creation of bacteria to breakdown toxic waste.
    D. The creation of organisms that are successfully being used as sources for alternative fuels.

Answer: D. The creation of organisms that are successfully being used as a source alternative fuels.

Although there is a push to genetically engineer organisms that will either create alternative fuels or be used as an alternative fuel source, this field is in it's infancy. There are multiple successful examples for each of the other posssible answers.

TEACHER CERTIFICATION STUDY GUIDE

41. **Which of the following is a way that cDNA cloning has not been used?**
   (Rigorous, Skill 2.3.8)

   A. To provide evidence for taxonomic organization.
   B. To study the mutations that lead to diseases such as hemophilia.
   C. To determine the structure of a protein.
   D. To understand methods of gene regulation.

Answer: C. To determine the structure of a protein.

Although cDNA cloning can be used to determine the amino acid sequence of a protein many other steps determine the final protein structure. For example, the folding of the protein, addition of other protein subunits, and/or modification by other proteins.

42. **A genetic engineering advancement in the medical field is...**
   (Easy, Skill 2.3.8)

   A. gene therapy.
   B. pesticides.
   C. degradation of harmful chemicals.
   D. antibiotics.

Answer: A. gene therapy.

Gene therapy is the introduction of a normal allele imto somatic cells in order to replace a defective gene. The medical field has had success in treating patients with a single enzyme deficiency disease. Gene therapy has allowed doctors and scientists to introduce a normal allele that provides the missing enzyme.

43. **The Law of Segregation defined by Mendel states...**
   (Average Rigor, Skill 3.1.1)

   A. when sex cells form, the two alleles that determine a trait will end up on different gametes.
   B. only one of two alleles is expressed in a heterozygous organism.
   C. the allele expressed is the dominant allele.
   D. alleles of one trait do not affect the inheritance of alleles on another chromosome.

Answer: A. when sex cells form, the two alleles that determine a trait will end up on different gametes.

The law of segregation states that the two alleles of each trait segregate to different gametes.

BIOLOGY

44. A child with type O blood has a father with type A blood and a mother with type B blood. The genotypes of the parents, respectively, would be which of the following?
(Average Rigor, Skill 3.1.1)

   A. AA and BO
   B. AO and BO
   C. AA and BB
   D. AO and OO

Answer: B. AO and BO

Type O blood has 2 recessive O genes. A child receives one allele from each parent; therefore, each parent in this example must have an O allele. The father has type A blood with a genotype of AO and the mother has type B blood with a genotype of BO.

45. Amniocentesis is…
   (Average Rigor, Skill 3.1.4)

   A. a non-invasive technique for detecting genetic disorders.
   B. a bacterial infection.
   C. extraction of amniotic fluid.
   D. removal of fetal tissue.

Answer: C. extraction of amniotic fluid.

Amniocentesis is a procedure in which a needle is inserted into the uterus to extract some of the amniotic fluid surrounding the fetus. Some genetic disorders can be detected by chemicals in the fluid.

46. A woman has Pearson Syndrome, a disease caused by a mutation in mitochondrial DNA. Which of the following individuals would you expect to see the disease in? *(Rigorous, Skill 3.1.5)*

   I   Her Daughter
   II Her Son
   III Her Daughter's son
   IV Her Son's daughter

   A. I, III
   B. I, II, III
   C. II, IV
   D. I, II, III, IV

Answer: B. I, II, III

Since mitochondrial DNA is passed through the maternal line, both of her children would be affected and the trait would continue to pass from her daughter to all of her children. Her son's children would recieve their mitochondrial DNA from their mother.

47. Which is not a possible effect of polyploidy?
    (Rigorous, Skill 3.1.5)

   A. More robust members of an animal species.
   B. The creation of cross species offspring.
   C. The creation of a new species.
   D. Cells that produce higher levels of desired proteins.

Answer: A. More robust members of an animal species.

While polyploidy often creates new plant species thereby yeilding more robust crops, it is likely to create nonviable animal offspring.

48.

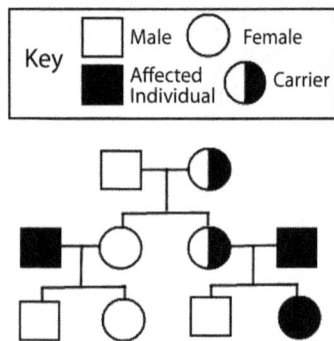

Based on the pedigree chart above, what term best describes the nature of the trait being mapped?
(Rigorous, Skill 3.1.5)

   A. Autosomal recessive.
   B. Sex-linked.
   C. Incomplete dominance.
   D. Co-dominance.

Answer: B. Sex-linked.

This chart would be a good example of color blindness, a sex-linked trait. If the trait had been autosomal recessive the last generation would all be carriers with the exception of the affected individual. In the case of traits that are incompletely dominant or co-dominant the tree would require additional notation.

49. Which aspect of science does not support evolution?
(Average Rigor, Skill 3.2.1)

   A. Comparative anatomy.
   B. Organic chemistry.
   C. Comparison of DNA among organisms.
   D. Analogous structures.

Answer: B. Organic chemistry.

Comparative anatomy is the comparison of anatomical characteristics between different species. This includes the study of homologous and analogous structures. The comparison of DNA between species is the best way to establish evolutionary relationships. Organic chemistry has nothing to do with evolution.

50. Fossils of the dinosaur genus Saurolophus have been found in both Western North America and Mongolia. What is the most likely explanation for these findings?
   (Rigorous, Skill 3.2.1)

   A. Convergent evolution on the two continents lead to two species of dinosaurs with sufficient similarity to be placed in the same genus.
   B. With the gaps in the fossil record, there are currently multiple competing theories to explain the presence of these fossils on two separate continents.
   C. A distant ancestor of the these dinosaurs evolved before these land masses became separated.
   D. Although Asia and North American were separate continents at the time, low sea levels made it possible for the dinosaurs to walk from one continent to the other.

Answer: D. Although Asia and North American were separate continents at the time, low sea levels made it possible for the dinosaurs to walk from one continent to the other.

Convergent evolution explains how different species develop similar traits but are classified in a different genus because of their greater differences. In the example provided, the fossil record indicates too many similarities, and thus the Saurolophus are a single genus of dinosaurs. (Koalas are an interesting example of convergent evolution. They are one of the few species to have distinct fingerprints like humans). Although there are gaps in the fossil record, there is one theory that is supported by significant evidence and to which most of the paleontologists ascribe. The last time the two land masses were part of the same continent (known as Laurasi) was approximately 40 million years before the existence of Saurolophus. The last and most likely possibility is that the dinosaurs walked from Asia to North America via the Bering Land Bridge (where the Bering Strait is now). The Bering Land bridge existed because large glaciers sequestered enough water to lower the level of the ocean. This lowering of ocean level was enough to expose what is now the ocean floor between Alaska and Siberia.

51. Which of the following best exemplifies the Theory of Inheritance of Acquired Traits?
(Average Rigor, Skill 3.2.2)

A. Giraffes need to reach higher for leaves to eat, so their necks stretch. The giraffe babies are then born with longer necks. Eventually, there are more long-necked giraffes in the population.
B. Giraffes with longer necks are able to reach more leaves, so they eat more and have more babies than other giraffes. Eventually, there are more long-necked giraffes in the population.
C. Giraffes want to reach higher for leaves to eat, so they release enzymes into their bloodstream, which in turn causes fetal development of longer-necked giraffes. Eventually, there are more long-necked giraffes in the population.
D. Giraffes with long necks are more attractive to other giraffes, so they get the best mating partners and produce more babies. Eventually, there are more long-necked giraffes in the population.

Answer: A. Giraffes need to reach higher for leaves to eat, so their necks stretch. The giraffe babies are then born with longer necks. Eventually, there are more long-necked giraffes in the population.

The theory of inheritance of acquired traits states that the offspring of an individual will benefit from the adaptations of the parent. The stretching of the neck thus leads to longer neck offspring. Answer b best exemplifies the theory of natural selection, where an outside factor affects the chance of an individual to live and reproduce, and thus pass on their genetic material to the next generation. There is no evidence of desire creating genetic or developmental change in a fetus. Additionally there is no evidence that giraffes select mates based on neck length, however if they did this would be an example of sexual selection, an aspect of natural selection.

52. Any change that affects the sequence of nucleotides in a gene is called a(n)...
    (Easy, Skill 3.2.2)

    A. deletion.
    B. polyploid.
    C. mutation.
    D. duplication.

Answer: C. mutation.
A mutation is an inheritable change in DNA. It may be an error in replication or a spontaneous rearrangement of one ore more segments of DNA. Deletion and duplication are types of mutations. Polyploidy is when an organism has more than two complete chromosome sets.

53. The fossil record is often cited as providing evidence for the theory of punctuated equilibrium. Which of the following examples can only be explained by punctuated equilibrium and not gradualism?
    (Rigorous, Skill 3.2.2)

    I   Coelacanth fish (once thought extinct) have remained relatively unchanged for millions of years.
    II The sudden apearance of a large number of different soft-bodied animals around 530 million years ago.
    III 10 million year old fossils and modern ginko plants are nearly identical.
    IV Fossils of Red Deer from the Island of Jersey show a six-fold decrease in body weight over the last 6000 years.

    A. I, III
    B. II, IV
    C. I, II, III, IV
    D. None of the above

Answer: D. None of the above

Gradualism and punctuated equilibrium are not mutually exclusive. Since we are talking in terms of geological time, a rapid change can be thought to occur over a period of 1,000 years to 100,000 years or more. Items I and III appear to demonstrate a state of stasis, however it is possible that some changes cannot be observed in the fossil record. Items II and IV appear to show a period of sudden change. In the case of item II, the change may have occurred over the course of a million years or more. In the case of item IV, 6000 years can conceivably include 3000 generations (red deer mature at age 2). Therefore, in both item II and IV, the apparent sudden change may have actually occurred gradually.

TEACHER CERTIFICATION STUDY GUIDE

54. Evolution occurs in...
(Easy, Skill 3.2.3)

- A. individuals.
- B. populations.
- C. organ systems.
- D. cells.

Answer: B. populations.

Evolution is a change in genotype over time. Gene frequencies shift and change from generation to generation. Populations evolve, not individuals.

55. Which of the following factors will affect the Hardy-Weinberg law of equilibrium, leading to evolutionary change?
(Average Rigor, Skill 3.2.3)

- A. No mutations.
- B. Non-random mating.
- C. No immigration or emigration.
- D. Large population.

Answer: B. Non-random mating.

There are five requirements to maintain Hardy-Weinberg equilibrium: no mutation, no selection pressures, an isolated population, a large population, and random mating.

56. **If a population is in Hardy-Weinberg equilibrium and the frequency of the recessive allele is 0.3, what percentage of the population would be expected to be heterozygous?**
    (Rigorous, Skill 3.2.3)

    A. 9%
    B. 49%
    C. 42%
    D. 21%

Answer: C. 42%

0.3 is the value of q. Therefore, $q^2 = 0.09$. According to the Hardy-Weinberg equation, $1 = p + q$.

$1 = p + 0.3$.
$p = 0.7$
$p^2 = 0.49$

Next, plug q^2 and p^2 into the equation $1 = p^2 + 2pq + q^2$.

$1 = 0.49 + 2pq + 0.09$ (where 2pq is the number of heterozygotes).
$1 = 0.58 + 2pq$
$2pq = 0.42$

Multiply by 100 to get the percent of heterozygotes, 42%.

57. **An animal choosing its mate because of attractive plumage or a strong mating call is an example of…**
    (Average Rigor, Skill 3.2.3)

    A. sexual selection.
    B. natural selection.
    C. mechanical isolation.
    D. linkage.

Answer: A. sexual selection.

Sexual selection, the act of choosing a mate, allows animals to have some choice in the genetic makeup of its offspring. The answer is (A).

TEACHER CERTIFICATION STUDY GUIDE

58. **Which of the following is an example of a phenotype that gives the organism an advantage in their home environment?**
    (Rigorous, Skill 3.2.3)

    A. The color of Pepper Moths in England
    B. Thornless roses in a nature perserve
    C. Albinism in naked mole rats
    D. The large thorax of a Mediterranean fruit fly

Answer: A. The color of Pepper Moths in England

Thornless roses are not naturally occuring and would not convey an advantage against natural predators. Albinism is not any more common in naked mole rats than other species and would not be advantageous in a subterrian environment. Thorax size in Mediterranean fruit flies has been linked to sexual selection, however sexual selection is not an environmental pressure. The Pepper Moth of England is the most often cited example of natural selection. A dramatic shift in color frequency occured during the industrial revolution. This color change occurred because moths camouflage on trees and during the industrial revolution soot changed the color of trees.

59. **Members of the same species:**
    (Easy, Skill 3.2.4)

    A. look identical.
    B. never change.
    C. reproduce successfully within their group.
    D. live in the same geographic location.

Answer: C. reproduce successfully within their group.

Species are defined by the ability to successfully reproduce with members of their own kind.

60. **Reproductive isolation results in...**
    (Average Rigor, Skill 3.2.4)

    A. extinction.
    B. migration.
    C. follilization.
    D. speciation.

Answer: D. speciation.

Reproductive isolation is caused by any factor that impedes two species from producing viable, fertile hybrids. Reproductive isolation of populations is the primary criterion for recognition of species status.

61. **The biological species concept applies to**
    (Average Rigor, Skill 3.2.4)

    A. asexual organisms.
    B. extinct organisms.
    C. sexual organisms.
    D. fossil organisms.

Answer: C. sexual organisms.

The biological species concept states that a species is a reproductive community of populations that occupy a specific niche in nature. It focuses on reproductive isolation of populations as the primary criterion for recognition of species status. The biological species concept does not apply to organisms that are completely asexual in their reproduction, fossil organisms, or distinctive populations that hybridize.

62. **The first cells that evolved on earth were probably of which type?**
    (Easy, Skill 3.2.6)

    A. Autotrophs.
    B. Eukaryotes.
    C. Heterotrophs.
    D. Prokaryotes.

Answer: D. Prokaryotes

Prokaryotes were first observed in the fossil record 3.5 billion years ago. Their ability to adapt to the environment allows them to thrive in a wide variety of habitats.

63. **All of the following gasses made up the primitive atmosphere except...**
    (Average Rigor, Skill 3.2.6)

    A. ammonia
    B. methane
    C. oxygen
    D. hydrogen

Answer: C. Oxygen

In the 1920's, Oparin and Haldane were the first to theorize that the primitive atmosphere was a reducing atmosphere with no oxygen. The atmosphere was rich in hydrogen, methane, water, and ammonia.

TEACHER CERTIFICATION STUDY GUIDE

64. Man's scientific name is Homo sapiens. Choose the proper classification beginning with kingdom and ending with order. *(Average Rigor, Skill 4.1.1)*

    A. Animalia, Vertebrata, Mammalia, Primate, Hominidae
    B. Animalia, Vertebrata, Chordata, Homo, sapiens
    C. Animalia, Chordata, Vertebrata, Mammalia, Primate
    D. Chordata, Vertebrata, Primate, Homo, sapiens

Answer: C. Animalia, Chordata, Vertebrata, Mammalia, Primate

The order of classification for humans is as follows: Kingdom, Animalia; Phylum, Chordata; Subphylum, Vertebrata; Class, Mammalia; Order, Primate; Family, Hominadae; Genus, Homo; Species, sapiens.

65. The two major ways to determine taxonomic classification are… *(Average Rigor, Skill 4.1.1)*

    A. evolution and phylogeny.
    B. reproductive success and evolution.
    C. phylogeny and morphology.
    D. size and color.

Answer: C. phylogeny and morphology.

Taxonomy is based on structure (morphology) and evolutionary relationships (phylogeny).

66. The scientific name Canis familiaris refers to the animal's… *(Easy, Skill 4.1.1)*

    A. kingdom and phylum names.
    B. genus and species names.
    C. class and species names.
    D. order and family names.

Answer: B. genus and species names.

Each species is scientifically known by a two-part name, a system called binomial nomenclature. The first word in the name is the genus and the second word is its specific epithet (species name).

BIOLOGY

67. **Thermoacidophiles are...**
    (Average Rigor, Skill 4.1.2)

    A. prokaryotes.
    B. eukaryotes.
    C. protists.
    D. archaea.

Answer: D. archaea.

Thermoacidophiles, methanogens, and halobacteria are members of the archaea group. They are as different from prokaryotes as prokaryotes are from eukaryotes.

68. **Protists are classified into major groups according to...**
    (Average Rigor, Skill 4.1.2)

    A. their method of obtaining nutrition.
    B. reproduction.
    C. metabolism.
    D. their form and function.

Answer: D. their form and function.

The extreme variation in protist classification reflects their diverse form, function, and life style. The protists are often grouped as algae (plant-like), protozoa (animal-like), or fungus-like based on their similarity of characteristics.

69. **All of the following are examples of a member of Kingdom Fungi except...**
    (Easy, Skill 4.1.2)

    A. mold.
    B. algae.
    C. mildew.
    D. mushrooms.

Answer: B. algae.

Mold, mildew, and mushrooms are all fungi. Brown and golden algae are members of the Kingdom Protista and green algae are members of the Plant Kingdom.

70. Which kingdom is comprised of organisms made of one cell with no nuclear membrane?
    (Easy, Skill 4.1.2)

    A. Monera
    B. Protista
    C. Fungi
    D. Algae

Answer: A. Monera

Monera is the only kingdom comprising unicellular organisms lacking a nucleus. Algae are classified as a protist. Algae may be uni- or multicellular and have a nucleus.

71. Within the Phylum Mollusca there are examples of both open and closed circulatory systems. Which of the following is a feature that is not common to both the open and closed cirulatory systems of molluscs?
    (Rigorous, Skill 4.1.2)

    A. Hemocoel
    B. Plasma
    C. Vessels
    D. Heart

Answer: A. Hemocoel

Hemocoel is the blood filled cavity that is present in animals with an open circulatory system. Unlike some other open circulatory systems, the molluscs have three blood vessels, two to bring blood from the lungs and one to push blood into the hemocoel.

72. Which of the following systems considers Archea (or Archeabacteria) as the most inclusive level of the taxonomic system?

    I   Three Domain System
    II Five Kingdom System
    III Six Kingdom System
    IV Eight Kingdom System
    (Rigorous, Skill 4.1.2)

    A. II, III
    B. I, IV
    C. I, III, IV
    D. I, II, III, IV

Answer: C. I, III, IV

In the five kingdom system the subkingdom Archaebacteriobionta is under the kingdom Monera.

73. Laboratory researchers distinguish and classify fungi from plants because the cell walls of fungi contain _____.
    (Rigorous, Skill 4.1.2)

    A. chitin.
    B. lignin.
    C. lipopolysaccharides.
    D. cellulose.

Answer: A. chitin.

All of the possible answers are compounds found in cell walls. Cellulose is found in the cell wall of all plants, while lignin is found only in the cell wall of vascular plants. Lipopolysaccharides are found in the cell wall of gram negative bacteria. Chitin is the only compound uniquely found in fungal cell walls.

74.

Identify the correct characteristics of the plant pictured above.
(Rigorous, Skill 4.2.1)

    A. seeded, non-vascular
    B. non-seeded, vascular
    C. non-seeded, non-vascular
    D. seeded, vascular

Answer: B. non-seeded, vascular

The picture above is of a fern, Division Pterophyta, which is a spore bearing vascular plant.

75. **Which of the following is a characteristic of a monocot?**
(Rigorous, Skill 4.2.1)

    A. parallel veins in leaves
    B. flower petals occur in multiples of 4 or 5
    C. two seed leafs
    D. vascular tissue absent from the stem

Answer: A. parallel veins in leaves

Monocots have one cotelydon, parallel veins in their leaves, and their flower petals are in multiples of threes. Dicots have flower petals in multiples of fours and fives.

76. **Spores are the reproductive mode for which of the following group of plants?**
   (Average Rigor, Skill 4.2.1)

   A. algae
   B. flowering plants
   C. conifers
   D. ferns

Answer: D. ferns

Ferns are non-seeded vascular plants. All plants in this group have spores and require water for reproduction. Algae, flowering plants, and conifers are not in this group of plants.

77.

Using the following taxonomic key identify the tree that this branch came from?

1 - Are the leaves PALMATELY COMPOUND (BLADES arranged like fingers on a hand)? – go to question 2
1 - Are the leaves PINNATELY COMPOUND (BLADES arranged like the vanes of a feather)? – go to question 3

2 - Are there usually 7 BLADES - Aesculus hippocastanum
2 - Are there usually 5 BLADES - Aesculus glabra

3 - Are there mostly 3-5 BLADES that are LOBED or coarsely toothed? - Acer negundo
3 - Are there mostly 5-13 BLADES with smooth or toothed edges? - Fraxinus Americana
(Rigorous)

    A. Aesculus hippocastanum
    B. Aesculus glabra
    C. Acer negundo
    D. Fraxinus Americana

Answer: C. Acer negundo

The leaves are pinnately compound, with 5 coarsly toothed leaves, leading to the answer: Acer negundo. The list below includes the scientific name and the common name for the all the plants listed above:
Aesculus hippocastanum (Horsechestnut)
Aesculus glabra (Ohio Buckeye)
Acer negundo (Boxelder, Ashleaf Maple)
Fraxinus Americana (White Ash)

78. **Which of the following is not a factor that affects the rate of both photosynthesis and respiration in plants?**
    (Average Rigor, Skill 4.2.3)

    A. the concentration of NADP and FAD
    B. the temperature
    C. the structure of plants
    D. the availability of different substrates

Answer: C. the structure of plants

The structure of the plants leaf affects its ability to absorb light which affects the rate of photosynthesis but not the rate of respiration.

79. **Oxygen is given off in the:**
    (Easy, Skill 4.2.3)

    A. light reaction of photosynthesis.
    B. dark reaction of photosynthesis.
    C. Krebs cycle.
    D. reduction of NAD^+ to NADH.

Answer: A. light reaction of photosynthesis.

The conversion of solar energy to chemical energy occurs in light reactions. As chlorophyll absorbs light, electrons are transferred and cause water to split, releasing oxygen as a waste product.

80. **The most ATP is generated through...**
    (Rigorous, Skill 4.2.3)

    A. fermentation.
    B. glycolysis.
    C. chemiosmosis.
    D. the Krebs cycle.

Answer: C. chemiosmosis.

The electron transport chain uses electrons to pump hydrogen ions across the mitochondrial membrane. This ion gradient is used to form ATP in a process called chemiosmosis. ATP is generated by the removal of hydrogen ions from NADH and $FADH_2$. This yields 34 ATP molecules.

TEACHER CERTIFICATION STUDY GUIDE

81. Which of the following is not employed by a young cactus to survive in an arid environment?
   (Rigorous, Skill 4.2.3)

   A. Stem as the principle site of photosynthesis.
   B. A deep root system to reach additional sources of groundwater.
   C. CAM cycle photosynthesis.
   D. Spherical growth form.

Answer: B. A deep root system to reach additional sources of groundwater.

A waxy sperical stem as the site of photosynthesis is an adaptation that limits water loss and allows for maximum water storage. CAM cycle photosynthesis allows for the plant to open its stomata at night thus limiting possible water loss due to evaporation. Some cacti will develop a taproot when it is necessary to stabilize the plant.

82. Oxygen created in photosynthesis comes from the breakdown of
   (Average Rigor, Skill 4.2.3)

   A. carbon dioxide.
   B. water.
   C. glucose.
   D. carbon monoxide.

Answer: B. water.

In photosynthesis, water is split; the hydrogen atoms are pulled to carbon dioxide which is taken in by the plant and ultimately reduced to make glucose. The oxygen from water is given off as a waste product.

83. A plant cell is placed in salt water. The resulting movement of water out of the cell is called…
   (Average Rigor, Skill 4.2.3)

   A. facilitated diffusion.
   B. diffusion.
   C. transpiration.
   D. osmosis.

Answer: B. diffusion.
Osmosis is simply the diffusion of water across a semi-permeable membrane. Water will diffuse out of the cell if less water is present on the outside than inside the cell.

BIOLOGY

84. **Double fertilization refers to which of the following?**
    (Average Rigor, Skill 4.2.4)

    A. two sperm fertilizing one egg
    B. fertilization of a plant by gametes from two separate plants
    C. two sperm enter the plant embryo sac; one sperm fertilizes the egg, the other forms the endosperm
    D. the production of non-identical twins through fertilization of two separate eggs

Answer: C. two sperm enter the plant embryo sac; one sperm fertilizes the egg, the other forms the endosperm

In angiosperms, double fertilization is when an ovum is fertilized by two sperm. One sperm produces the new plant and the other forms the food supply for the developing plant (endosperm).

85. **The process in which pollen grains are released from the anthers is called:**
    (Easy, Skill 4.2.4)

    A. pollination.
    B. fertilization.
    C. blooming.
    D. dispersal.

Answer: A. pollination.

Pollen grains are released from the anthers during pollination and are carried by the wind and animals to the carpels.

86. **Which of the following is the correct order of the stages of plant development from egg to adult plant?**
    (Average Rigor, Skill 4.2.4)

    A. morphogenesis, growth, and cellular differentiation
    B. cell differentiation, growth, and morphogenesis
    C. growth, morphogenesis, and cellular differentiation
    D. growth, cellular differentiation, and morphogenesis

Answer: C. growth, morphogenesis, and cellular differentiation

The development of the egg to form a plant occurs in three stages: growth; morphogenesis; and cellular differentiation, the acquisition of a cell's specific structure and function.

TEACHER CERTIFICATION STUDY GUIDE

87. In angiosperms, the food for the developing plant is found in which of the following structures?
    (Average Rigor, Skill 4.2.4)

    A. ovule
    B. endosperm
    C. male gametophyte
    D. cotyledon

Answer: B. Endosperm

The endosperm is a product of double fertilization. It is the food supply for the developing plant.

88. In a plant cell, telophase is described as...
    (Rigorous, Skill 4.2.4)

    A. the time of chromosome doubling.
    B. cell plate formation.
    C. the time when crossing over occurs.
    D. cleavage furrow formation.

Answer: B. cell plate formation.

In a plant cell, a cell plate forms during telophase. In an animal cell, a cleavage furrow forms during telophase.

89. Which of the following is a disadvantage of budding compared to sexual reproduction?
    (Rigorous, Skill 4.2.4)

    A. limited number of offspring
    B. inefficient
    C. limited genetic diversity
    D. expensive to the parent organism

Answer: C. limited genetic diversity

Budding results in an identical copy of the parent, therefore the genetic diversity of the individuals in an area is very limited. It is possible for an organism that reproduces by budding to produce a much larger number of offspring than an organism reproducing by sexual reproduction.

BIOLOGY

90. **Which phylum accounts for 85% of all animal species?**
    (Easy, Skill 4.3.1)

    A. Nematoda
    B. Chordata
    C. Arthropoda
    D. Cnidaria

Answer: C. Arthropoda

The arthropoda phylum consists of insects, crustaceans, and spiders. They are the largest group in the animal kingdom.

91. **Fats are broken down by which substance?**
    (Average Rigor, Skill 4.3.2)

    A. bile produced in the gall bladder
    B. lipase produced in the gall bladder
    C. glucagons produced in the liver
    D. bile produced in the liver

Answer: D. bile produced in the liver

The liver produces bile, which breaks down and emulsifies fatty acids.

92. **A boy had the chicken pox as a baby. He will most likely not get this disease again because of**
    (Average Rigor, Skill 4.3.2)

    A. passive immunity
    B. vaccination.
    C. antibiotics.
    D. active immunity.

Answer: D. active immunity.

Active immunity develops after recovery from an infectious disease, such as the chicken pox, or after vaccination. Passive immunity may be passed from one individual to another (from mother to nursing child).

BIOLOGY

TEACHER CERTIFICATION STUDY GUIDE

93. Movement is possible by the action of muscles pulling on *(Average Rigor, Skill 4.3.2)*

    A. skin.
    B. bones.
    C. joints.
    D. ligaments.

Answer: B. bones.

Skeletal muscles are attached to bones and are responsible for their movement.

94. Hormones are essential for the regulation of reproduction. What organ is responsible for the release of hormones for sexual maturity? *(Average Rigor, Skill 4.3.2)*

    A. pituitary gland
    B. hypothalamus
    C. pancreas
    D. thyroid gland

Answer: B. Hypothalamus

The hypothalamus begins secreting hormones that help mature the reproductive system and stimulate development of secondary sex characteristics.

95. Which of the following compounds is not needed for skeletal muscle contraction to occur? *(Rigorous, Skill 4.3.2)*

    A. glucose
    B. sodium
    C. acetylcholine
    D. Adenosine 5'-triphosphate

Answer: A. Glucose

Although glucose is necessary to generate ATP (Adenosine 5'-triphosphate) it is not directly involved in muscular contractions. Acetylocholine is the neurotransmitter that intiates muscle contraction. Sodium plays an essential part in creating an action potential. Lastly, ATP provides the energy for contraction.

96. **Which of the following hormones is most involved in the process of osmoregulation?**
    (Rigorous, Skill 4.3.2)

    A. Antidiuretic Hormone.
    B. Melatonin.
    C. Calcitonin.
    D. Gulcagon.

Answer: A. Antidiuretic Hormone.

The mechanism through which the body controls water concentration and various soluble materials is called osmoregulation. Antidiuretic Hormone (ADH) regulates the kidneys' reabsorption of water and directly affects the amount of water in the body. A failure to produce ADH can cause an individual to die from dehydration within a matter of hours. Calcitonin controls the removal of calcium from the blood. Glucagon, like insulin, controls the amount of glucose in the blood. Like ADH melatonin plays a role in homeostasis, by regulating body rhythms.

97. **Capillaries come into contact with a large surface of both the kidneys and the lungs, especially in relation to the volume of these organs. Which of the following is not consistent with both organs and their contact with capillaries.**
    (Rigorous, Skill 4.3.2)

    A. small specialized sections of each organ contact capillaries
    B. A large branching system of tubes within the organ
    C. A large source of blood that is quickly divided into capillaries
    D. A sack that contains a capillary network

Answer: D. A sack that contains a capillary network

The Bowmen's capsule of the kidneys can be described as a sack that contains a capillary network. The alveoli of the lungs are sacks, however the capillaries are on the outside of the alveoli.

98. Which of the following substances in unlikely to cause negative consequences if over-ingested?
    (Rigorous, Skill 4.3.2)

    A. essential fatty acids
    B. essential minerals
    C. essential water-insoluble vitamins
    D. essential water-soluble vitamins

Answer: D. essential water-soluble vitamins

Water-soluble vitamins are often removed in the filtration process by the kidneys and thus rarely build up to dangerous levels. Too many fatty acids can lead to obesity and other health problems. Excessive minerals can lead to a variety of different conditions, depending on the mineral ingested. Water-insoluble vitamins are usually stored in fatty tissues and thus are not flushed from the body. Therefore, water-insoluble vitamins can build up reaching dangerous levels.

99. If someone were experiencing unexplained changes in body temperature, hunger, and circadian rythyms, which of the following structures would most likely be the cause of these problems?
    (Rigorous, Skill 4.3.2)

    A. hypothalamus
    B. central nervous system
    C. pineal gland
    D. basal ganglia

Answer: A. hypothalamus

The pineal gland releases melatonin, which has been linked to sleep/wake patterns. The basal ganglia and central nervous system are structures regulating nerve impulses. Only the hypothalamus is responsible for regulating body temperature, hunger, and sleep/wake cycles.

100. **Which is the correct sequence of embryonic development in a frog?**
     (Average Rigor, Skill 4.3.3)

    A. cleavage – blastula – gastrula
    B. cleavage – gastrula – blastula
    C. blastula – cleavage – gastrula
    D. gastrula – blastula – cleavage

Answer: A. cleavage – blastula – gastrula

Animals go through several stages of development after egg fertilization. The first step is cleavage which continues until the egg becomes a blastula. The blastula is a hollow ball of undifferentiated cells. Gastrulation follows and is the stage in which tissue differentiates into separate germ layers: the endoderm, mesoderm, and ectoderm.

101. **Fertilization in humans usually occurs in the:**
     (Easy, Skill 4.3.3)

    A. cervix.
    B. ovary.
    C. fallopian tubes.
    D. vagina.

Answer: C. fallopian tubes.

Fertilization of the egg by the sperm normally occurs in the fallopian tube. The fertilized egg then implants in the uterine lining for development.

102. **Which of the following has no relation to female sexual maturity?**
     (Rigorous, Skill 4.3.3)

    A. thyroxine
    B. estrogen
    C. testosterone
    D. luteinizing hormone

Answer: A. thyroxine

Thyroxine is a hormone associated with the regulation of the body's metabolism, and is not related to the maturation process. Luteinizng hormone stimulates cells in both the testes and ovaries. Estrogen is the hormone most frequently associated with female development. Studies have indicated that levels of testosterone increase as a female goes through puberty and drop off after she reaches her sexual peak.

TEACHER CERTIFICATION STUDY GUIDE

103. **In the growth of a population, initially the increase is exponential until carrying capacity is reached. This is represented by a(n):**
(Average Rigor, Skill 5.1)

   A. S curve.
   B. J curve.
   C. M curve.
   D. L curve.

Answer: A. S curve.

An exponentially growing population starts off with little change and then rapidly increases. The graphic representation of this growth curve has the appearance of a "J". However, as the carrying capacity of the growing population is reached, the growth rate begins to slow down and level off. The graphic representation of this growth curve has the appearance of an "S".

104. **All of the following are density dependent factors that affect a population except**
(Rigorous, Skill 5.1)

   A. disease.
   B. drought.
   C. predation.
   D. migration.

Answer: B. drought.

Although drought would affect the amount of food available to a population (which creates a density dependent factor), the drought itself would occur regardless of population size, and is thus density independent. Disease and migration tend to occur more frequently in crowded populations. The amount of prey and predators would affect the number of individuals in a population.

105. **Which of the following is not an example of dynamic equilibrium?**
    (Rigorous, Skill 5.1)

    A. a stable population
    B. a symbiotic pair of organisms
    C. osmoregulation
    D. maintaining head position while walking

Answer: D. maintaining head position while walking

Maintinaing head position while walking is a case of static equilibrium, a state where things do not change. In a stable population birth and death rates must balance. In a symbiotic pair, the contributions of each organism must balance, or the relationship becomes parasitic. Osmoregulation balances the body's need for water and the dissolved compounds within it.

106. **All of the following are density independent factors that affect a population except**
    (Average Rigor, Skill 5.1)

    A. temperature.
    B. rainfall.
    C. predation.
    D. soil nutrients.

Answer: C. predation

As a population increases, the competition for resources intensifies and the growth rate declines. This is a density-dependent factor. An example of this would be predation. Density-independent factors affect the population regardless of its size. Examples of density-independent factors are rainfall, temperature, and soil nutrients.

107. **If the niches of two species overlap, what usually results?**
    (Easy, Skill 5.2)

    A. a symbiotic relationship
    B. cooperation
    C. competition
    D. a new species

Answer: C. competition

Two species that occupy the same habitat or eat the same food are said to be in competition with one another.

108. **Primary succession occurs after...**
    (Average Rigor, Skill 5.2)

    A. nutrient enrichment.
    B. a forest fire.
    C. exposure of a bare rock after the water table permanently recedes.
    D. a housing development is built.

Answer: C. exposure of a bare rock after the water table permanently recedes.

Primary succession occurs where life never existed before, such as flooded areas or a new volcanic island. It is only after the water recedes that the rock is able to support new life.

109. **A clownfish is protected by the sea anemone's tentacles. In turn, the anemone receives uneaten food from the clownfish. This is an example of...**
    (Easy, Skill 5.2)

    A. mutualism.
    B. parasitism.
    C. commensalism.
    D. competition.

Answer: A. mutualism.

Neither the clownfish nor the anemone cause harmful effects towards one another and they both benefit from their relationship. Mutualism is when two species that occupy a similar space benefit from their relationship.

TEACHER CERTIFICATION STUDY GUIDE

110. Which of the following are reasons to maintain biological diversity?

    I. Consumer product development.
    II. Stability of the environment.
    III. Habitability of our planet.
    IV. Cultural diversity.
    (Rigorous, Skill 5.2)

    A. I and III
    B. II and III
    C. I, II, and III
    D. I, II, III, and IV

Answer: D. I, II, III, and IV

Biological diversity refers to the extraordinary variety of living things and ecological communities throughout the world. Maintaining biological diversity is important for many reasons. First, we derive many consumer products used by humans from living organisms in nature. Second, the stability and habitability of the environment depends on the varied contributions of many different organisms. Finally, the cultural traditions of human populations depend on the diversity of the natural world. The answer is (D).

111. Which of the following is not an abiotic factor?
    (Easy, Skill 5.2)

    A. temperature
    B. rainfall
    C. soil quality
    D. bacteria

Answer: D. bacteria

Abiotic factors are the non-living aspects of an ecosystem. Bacteria is an example of a biotic factor, a living thing.

112. **In the nitrogen cycle, decomposers are responsible for which process?**
    (Rigorous, Skill 5.3)

    A. Nitrogen fixing
    B. Nitrification
    C. Ammonification
    D. Assimilation

Answer: C. Ammonification

Nitrogen fixing and nitrification are primarily handled by bacteria (although inorganic processes augment this somewhat). Assimilation is a process performed by plants. Decomposers ammonify thus preparing nitrogen compounds for bacteria.

113. **Which biome is the most prevalent on Earth?**
    (Average Rigor, Skill 5.3)

    A. marine
    B. desert
    C. savanna
    D. tundra

Answer: A. Marine

The marine biome covers 75% of the Earth. This biome is organized by the depth of water.

114. **Which term is not associated with the water cycle?**
    (Easy, Skill 5.3)

    A. precipitation
    B. transpiration
    C. fixation
    D. evaporation

Answer: C. fixation

Water is recycled through the processes of evaporation and precipitation. Transpiration is the evaporation of water from leaves. Fixation is not associated with the water cycle.

TEACHER CERTIFICATION STUDY GUIDE

115. Which of the following terms does not describe a way that the human race negatively impacts the biosphere?
    (Rigorous, Skill 5.3)

    A. biological magnification
    B. pollution
    C. carrying capacity
    D. simplification of the food web

Answer: C. carrying capacity

Most people recognize the harmful effects of pollution, especially global warming. Pollution, and the regular use of pesticides and herbicides introduces toxins into the food web, biological magnification refers to the increasing concentration of these toxins as you move up the food web. Simplification of the food web has to do with small variety farming crops replacing large habitats, and thus shrinking or destroying some ecosystems. Carrying capacity on the other hand is simply a term for the amount of life a certain habitat can sustain, it is term independent of human action, so the answer is (C).

116. Which biogeochemical cycle plays the smallest part in photosynthesis or cellular respiration?
    (Rigorous, Skill 5.3)

    A. Hydrogen cycle
    B. Phosphorous cycle
    C. Sulfur cycle
    D. Nitrogen cycle

Answer: C. Sulfur cycle

Although sulfur is used in a small part of the proteins used in photosynthesis and respiration, only in some photosynthesizing bacteria (not cyanobacteria) does sulfur play a significant role. Hydrogen is readily passed through almost every stage of respiration and photosynthesis. Many of the energy storing molecules (including ATP) use phosphorous as a principle component. Nitrogen, in addition to being a significant component of proteins that assist in these processes, is also a significant component of NADP.

117. Genetic engineering is beneficial to agriculture in many ways. Which of the following is not an advantage of genetic engineering to agriculture?
    (Average Rigor, Skill 6.1)

    A. Use of bovine growth hormone to increase milk production.
    B. Development of crops resistant to herbicides.
    C. Development of microorganisms to breakdown toxic substances into harmless compounds.
    D. Genetic vaccination of plants against viral attack.

Answer: C. The development of microorganisms to breakdown toxic substances into harmless compounds.

All of the answers are actual results of genetic engineering, however only answer (C) has not been used for agricultural purposes. These microorganisms have however been used at toxic waste sites and oil spills.

118. The biggest problem that humans have created in our environment is:
    (Average Rigor, Skill 6.2)

    A. Nutrient depletion and excess.
    B. Global warming.
    C. Population overgrowth.
    D. Ozone depletion.

Answer: A. Nutrient depletion and excess.

Human activity affects parts of nutrient cycles by removing nutrients from one part of the biosphere and adding them to another. This results in nutrient depletion in one area and nutrient excess in another. This affects water systems, crops, wildlife, and humans. The answer is (A).

TEACHER CERTIFICATION STUDY GUIDE

119. **The demand for genetically enhanced crops has increased in recent years. Which of the following is not a reason for this increased demand?**
(Easy, Skill 6.2)

   A. Fuel sources
   B. Increased growth
   C. Insect resistance
   D. Better-looking produce

Answer: D. Better-looking produce

Genetically enhanced crops are being developed for utilization as fuel sources, as well as for an increased production yield. Insect resistance eliminates the need for pesticides. While there may be some farmers crossing crops to make prettier watermelons, this is not a primary reason for the increased demand. The answer is (D).

120. **Stewardship is the responsible management of resources. We must regulate our actions to do which of the following about environmental degradation?**
(Average Rigor, Skill 6.3)

   A. Prevent it.
   B. Reduce it.
   C. Mitigate it.
   D. All of the above.

Answer: D. All of the above.

Stewardship requires the regulation of human activity to prevent, reduce, and mitigate environmental degradation. An important aspect of stewardship is the preservation of resources and ecosystems for future human generations. Therefore, the answer is (D).

TEACHER CERTIFICATION STUDY GUIDE

121. The three main concerns in nonrenewable resource management are conservation, environmental mitigation, and _____.
    (Rigorous, Skill 6.3)

    A. preservation
    B. extraction
    C. allocation
    D. sustainability

Answer: C. allocation

The main concerns in nonrenewable resource management are conservation, allocation, and environmental mitigation. Policy makers, corporations, and governments must determine how to use and distribute scarce resources. Decision makers balance the immediate demand for resources with the need for resources in the future. This determination is often a cause of conflict and disagreement. Finally, scientists attempt to minimize and mitigate the environmental damage caused by resource extraction. The answer is (C).

122. Which of the following limit the development of technological design ideas and solutions?

    I. monetary cost
    II. time
    III. laws of nature
    IV. governmental regulation
    (Average Rigor, Skill 6.4)

    A. I and II
    B. I, II, and IV
    C. II and III
    D. I, II, and III

Answer: D. I, II, and III

Technology cannot work against the laws of nature. Technological design solutions must work within the framework of the natural world. Monetary cost and time constraints also limit the development of new technologies. Governmental regulation, while present in many sciences, cannot regulate the formation of new ideas or design solutions. The answer, then, is D: I, II, and III.

123. **Which of the following is the least ethical choice for a school laboratory activity?**
    (Rigorous, Skill 6.4)

    A. Dissection of a donated cadaver.
    B. Dissection of a preserved fetal pig.
    C. Measuring the skeletal remains of birds.
    D. Pithing a frog to watch the circulatory system.

Answer: D. Pithing a frog to watch the circulatory system.

Scientific and societal ethics make choosing experiments in today's science classroom difficult. It is possible to ethically perform choices (A), (B), or (C), if due care is taken. (Note that students will need significant assistance and maturity to perform these experiments, and that due care also means attending to all legal paperwork that might be necessary.) However, modern practice precludes pithing animals (causing partial brain death while allowing some systems to function), as inhumane. Therefore, the answer to this question is (D).

www.ingramcontent.com/pod-product-compliance
Lightning Source LLC
Chambersburg PA
CBHW062127160426
43191CB00013B/2217